Electronics Formulas Pocket Reference

Electronics Formulas Pocket Reference

Stan Gibilisco

Editor in Chief

McGraw-Hill, Inc.
New York San Francisco Washington, D.C. Auckland Bogotá
Caracas Lisbon London Madrid Mexico City Milan
Montreal New Delhi San Juan Singapore
Sydney Tokyo Toronto

Library of Congress Cataloging-in-Publication Data
Electronics formulas pocket reference / Stan Gibilisco, editor-in-chief.
   p. cm.
  Includes bibliographical references and index.
  ISBN 0-07-135316-X
  1. Electronics—Handbooks, manuals, etc.   I. Gibilisco, Stan.
TK7825 .E365 2000
621.381′02′12—dc21   99-056279
                                                                                                    CIP

McGraw-Hill

A Division of The McGraw-Hill Companies

Copyright © 2000 by The McGraw-Hill Companies, Inc. All rights reserved. Printed in the United States of America. Except as permitted under the United States Copyright Act of 1976, no part of this publication may be reproduced or distributed in any form or by any means, or stored in a data base or retrieval system, without the prior written permission of the publisher.

1 2 3 4 5 6 7 8 9 0   DOC/DOC   9 0 9 8 7 6 5 4 3 2 1 0 9

ISBN 0-07-135316-X

The sponsoring editor for this book was Scott Grillo, the editing supervisor was Andrew Yoder and the production supervisor was Pamela A. Pelton.

Printed and bound by R. R. Donnelley & Sons Company.

McGraw-Hill books are available at special quantity discounts to use as premiums and sales promotions, or for use in corporate training programs. For more information, please write to the Director of Special Sales, McGraw-Hill, 11 West 19th Street, New York, NY 10011. Or contact your local bookstore.

Information in this book has been obtained by the publisher from sources believed to be reliable. However, neither the publisher nor the authors guarantee the accuracy or completeness of any information published herein. Neither the publisher nor the authors shall be responsible for any errors, omissions, or damages arising out of use of this information. This work is published with the understanding that the publisher and authors are supplying information but are not attempting to render professional services in any way, shape or form. If such services are required, the assistance of an appropriate professional should be sought.

*To Tim, Tony, and Samuel
from Uncle Stan*

Contents

Acknowledgments xiii

Preface xv

Chapter 1. Fundamental Units   1

  The SI System   1
  Electrical Units   4
  Magnetic Units   12
  Gain and Loss   14
  Miscellaneous Units   16

Chapter 2. Conversions and Constants   21

  Prefix Multipliers   22
  Alternative Unit Systems   22

viii   Contents

    SI Unit Conversions   22
    Electrical Unit Conversions   28
    Magnetic Unit Conversions   28
    Miscellaneous Unit Conversions   35
    Constants   35

Chapter 3. Mathematical Notation   43

    Greek Alphabet   43
    General Symbols   48
    Subscripts and Superscripts   54
    Scientific Notation   56
    Significant Figures   60
    Precedence of Operations   63

Chapter 4. Algebra and Trigonometry   65

    Theorems in Algebra   65
    Coordinate Systems   67
    Trigonometry   77
    Logarithms   79

Chapter 5. Sequences and Series   85

    Definitions   85
    Basic Series   89
    More Sophisticated Series   91
    Trigonometric Series   93
    Logarithmic and Exponential Series   93
    Trigonometric/Exponential Formulas   95

Contents ix

Chapter 6. Sets, Functions, and Vectors — 97

   Sets — 97
   Functions — 100
   Vectors — 108

Chapter 7. Differentiation and Integration — 115

   Derivatives — 115
   Integrals — 121

Chapter 8. Direct Current — 135

   DC Charge — 135
   DC Amperage — 136
   DC Voltage — 140
   DC Resistance — 144
   DC Power — 146
   DC Energy — 147

Chapter 9. Alternating Current — 151

   Frequency and Phase — 151
   AC Amplitude Expressions — 154
   Complex Numbers — 158
   Impedance — 160
   Admittance — 162
   AC Amperage — 166
   AC Voltage — 168
   AC Power — 170
   AC Energy — 172

x   Contents

Chapter 10. Magnetism and Transformers — 175

- Reluctance — 175
- Basic Formulas — 177
- Induced Voltage — 179
- Transformers — 180
- Losses in Transformers and Inductors — 183

Chapter 11. Digital Electronics — 187

- Numbering Systems — 187
- Basic Binary Operations — 193
- Secondary Binary Operations — 195
- Logic Gates — 196
- Boolean Theorems — 199
- Flip-flops — 201

Chapter 12. Resonance, Filters, and Noise — 205

- Resonant Frequency — 205
- Lowpass Filters — 208
- Highpass Filters — 213
- Bandpass Filters — 217
- Bandstop Filters — 222
- Noise — 227

Chapter 13. Semiconductors — 231

- Diodes — 231
- Bipolar Transistors — 235
- Field-Effect Transistors — 245

Chapter 14. Electron Tubes — 251

- Basic Behavior — 252
- Parameters — 253
- Circuit Formulas — 258
- Power Formulas — 262

Chapter 15. Electromagnetic Waves and Antenna Systems — 267

- Electromagnetic Fields — 267
- RF Transmission Lines — 270
- Antennas — 280

Chapter 16. Measurement — 287

- Bridge Circuits — 287
- Null Networks — 295
- Error and Interpolation — 300

Bibliography 303
Index 305

Acknowledgments

Illustrations in this book were generated with CorelDRAW. Some clip art is courtesy of Corel Corporation, 1600 Carling Avenue, Ottawa, Ontario, Canada K1Z 8R7.

Preface

This book is a quick reference source of formulas, units, constants, symbols, and conversion factors for use by electronics technicians, engineers, hobbyists, and students. Every effort has been made to arrange the material in a logical manner, and to portray the information in concise but complete terms.

Suggestions for future editions are welcome. I can be reached by e-mail from links to my Web site at:

http://members.aol.com/stangib

Stan Gibilisco
Editor in Chief

Chapter 1

Fundamental Units

This chapter contains definitions of basic units encountered in electronics and related disciplines.

The SI System

The Standard International (SI) System of Units, formerly called the *Meter/Kilogram/Second (MKS) System*, defines seven quantities manifested in nature. See Chapter 2 for conversions to and from other units.

Displacement

One *meter* (1 m) is equivalent to 1.65076373×10^6 wavelengths in a vacuum of the radiation corresponding to the transition between the two levels of the krypton-86 atom. Originally defined as 10^{-7} of the way from the North Geographic Pole to the equator, as measured over the surface of the earth. Displacement is represented in equations by the italicized lowercase letters d or s.

Mass

One *kilogram* (1 kg) is the mass of 1000 cubic centimeters (1.000×10^3 cm^3) of pure liquid water at the temperature of its greatest density (approximately 281 degrees Kelvin). Mass is represented in equations by the italicized lowercase letter m.

Time

One *second* (1 s) is $1/86{,}400 = 1.1574 \times 10^{-5}$ part of a solar day. Also defined as the time required for a beam of visible light to propagate over a distance of 2.99792×10^8 m in a vacuum. Time is represented in equations by the italicized lowercase letter t.

Temperature

One *degree Kelvin* (1°K) is 3.66086×10^{-3} part of the difference between absolute zero (the absence of all heat) and the freezing point of pure water at standard atmospheric temperature and pressure. Temperature is represented in equations by the italicized uppercase letter T.

Electric current

One *ampere* (1 A) represents the movement of 6.24×10^{18} charge carriers (usually electrons) past a specific fixed point in an electrical conductor over a time span of 1 s. Current is represented in equations by the italicized uppercase letter I.

Luminous intensity

One *candela* (1 cd) represents the radiation from a surface area of 1.667×10^{-6} m^2 of a blackbody at the solidification temperature of pure platinum. Luminous intensity is represented in equations by the italicized uppercase letters B, F, I, or L.

Material quantity

One *mole* (1 mol) is the number of atoms in precisely 0.012 kg of carbon-12. This is approximately 6.022169×10^{23}, also known as *Avogadro's number*. Material quantity is represented in equations by the italicized uppercase letter N.

4 Chapter One

Electrical Units

Electrical units are defined for many quantities and phenomena. Standard units, derived from SI basic units, are defined here. For conversions to and from other units representing these properties, see Chapter 2.

Unit electric charge

The *unit electric charge* is the charge contained in a single electron. This charge is also contained in a hole (electron absence within an atom), a proton, a positron, and an anti-proton. Charge quantity, in terms of unit electric charges, is represented in equations by the italicized lowercase letter e.

Electric charge quantity

The standard unit of electric charge quantity is the *coulomb* (C), which is the total charge contained in 6.24×10^{18} electrons. Charge quantity is represented in equations by the italicized uppercase Q or lowercase q.

Energy

The standard SI unit of energy is the *joule* (J). Mathematically, it is expressed in terms of unit mass multiplied by unit distance squared per unit time squared:

$$1 \text{ J} = 1 \text{ kg} \times \text{m}^2/\text{s}^2$$

Energy is represented in equations by the italicized uppercase letter E. Occasionally, it is represented by the italicized uppercase H, T, or V.

Electromotive force

The standard unit of electromotive force (EMF), also called *electric potential* or *potential difference*, is the *volt* (V). It is equivalent to 1 J/C. Electromotive force is represented in equations by the italicized uppercase E or V.

Resistance

The standard unit of resistance is the *ohm* (Ω). It is the resistance that results in 1 A of electric current with an applied EMF of 1 V. Resistance is represented in equations by the italicized uppercase letter R.

Resistivity

The standard unit of resistivity is the *ohm-meter* ($\Omega \times \text{m}$). If a length of material measuring 1 m carries 1 A of current when a potential difference of 1 V is applied, then it has a resistivity of 1 Ω × m. Resistivity is represented in equations by the italicized lowercase Greek letter ρ.

Conductance

The standard unit of conductance is the *siemens* (S), formerly called the *mho*. Mathematically, conductance is the reciprocal of resistance. Conductance is represented in equations by the italicized uppercase letter G. If R is the resistance of a component in ohms, and G is the conductance of the component in siemens, then:

$$G = 1/R$$

and

$$R = 1/G$$

Conductivity

The standard unit of conductivity is the *siemens per meter* (S/m). If a length of material measuring 1 m carries 1 A of current when a potential difference of 1 V is applied, then it has a conductivity of 1 S/m. Conductivity is represented in equations by the italicized lowercase Greek letter σ.

Power

The standard unit of power is the *watt* (W), equivalent to 1 J/s. Power is represented in equations by the italicized uppercase P or W. In electrical and electronic circuits containing no reactance, if P is the power in watts, E is the voltage in volts, I is the current in amperes, and R is the resistance in ohms, then the following holds:

$$P = E \times I = I^2 \times R = E^2/R$$

Period

The standard unit of alternating-current (AC) cycle period is the *second* (s). This is a large unit in practice; typical signals have periods on the order of thousandths, millionths, billionths, or trillionths of a second. Period is represented in equations by the italicized uppercase letter T.

Frequency

The standard unit of frequency is the *hertz* (Hz), formerly called the *cycle per second* (cps). This is a small unit in practice; typical signals have frequencies on the order of thousands, millions, billions, or trillions of hertz. Frequency, which is the mathematical reciprocal of period, is represented in equations by the italicized lowercase letter f or the italicized lowercase Greek letter ν. If T is the period of a wave disturbance in seconds, then the frequency (f, in hertz) is given by:

$$f = 1/T$$

Capacitance

The standard unit of capacitance is the *farad* (F), which is equal to 1 C/V. This is a large unit in practice. In electronic circuits, most values of capacitance are on the order of millionths, billionths, or trillionths of a farad. Capacitance is represented in equations by the italicized uppercase letter C.

Inductance

The standard unit of inductance is the *henry* (H), which is equal to 1 V × s/A. This is a large unit in practice. In electronic circuits, most values of inductance are on the order of thousandths or millionths of a henry. Inductance is represented in equations by the italicized uppercase letter L.

Reactance

The standard unit of reactance is the *ohm* (Ω). Reactance is represented in equations by the italicized uppercase letter X. It can be positive (inductive) and symbolized by X_L, or negative (capacitive) and symbolized by X_C. Reactance is dependent on frequency. For the following formulas, f represents frequency in hertz, L represents inductance in henrys, and C represents capacitance in farads:

$$X_L = 2 \times \pi \times f \times L$$
$$X_C = -1/(2 \times \pi \times f \times C)$$

Complex impedance

In the determination of complex impedance, there are two components: resistance (R) and reactance (X). The reactive component is multiplied by the unit imaginary number, known as the *j operator*. Mathematically, j is equal to the positive square root of -1, so:

$$j = (-1)^{1/2}$$
$$j^2 = j \times j = -1$$
$$j^3 = j^2 \times j = -1 \times j = -j$$
$$j^4 = j^3 \times j = -j \times j = 1$$

For powers of j beyond 4, the cycle repeats. So, in general, for integers $n > 4$:

$$j^n = j^{(n-4)}$$

Let Z represent complex impedance, R represent resistance, and X represent reactance (either inductive or capacitive). Then:

$$Z = R + jX$$

Absolute-value impedance

Complex impedance can be represented as a vector in a rectangular coordinate plane, where resistance is plotted on the abscissa (horizontal axis) and reactance is plotted on the ordinate (vertical axis). The length of this vector is called the *absolute-value impedance*, symbolized by the italicized uppercase Z and expressed in ohms. This impedance is usually discussed only when $jX = 0$; that is, when the impedance is a pure resistance ($Z = R$). In the broad sense, if Z is the absolute-value impedance, then

$$Z = (R^2 + X^2)^{1/2}$$

In theory, infinitely many combinations of R and X can result in a given absolute-value impedance Z.

Electric field strength

The standard unit of electric field strength is the *volt per meter* (V/m). An electric field of 1 V/m is represented by a potential difference of 1 V existing between two points displaced by 1 m. Electric field strength is represented in equations by the italicized uppercase letter E.

Electromagnetic field strength

The standard unit of electromagnetic (EM) field strength is the *watt per square meter* (W/m^2). An EM field of 1 W/m^2 is represented by 1 W of power impinging perpendicularly on a flat surface whose area is 1 m^2.

Electric susceptibility

The standard unit of electric susceptibility is the *coulomb per volt-meter*, abbreviated C/(Vm). This quantity is represented in equations by the italicized lowercase Greek letter η.

Permittivity

The standard unit of permittivity is the *farad per meter* (F/m). Permittivity is represented in equations by the italicized lowercase Greek letter ε.

Charge-carrier mobility

The standard unit of charge-carrier mobility, also called *carrier mobility* or simply *mobility*, is the *meter squared per volt-second*, abbreviated m²/(V × s). Mobility is represented in equations by the italicized lowercase Greek letter μ.

Magnetic Units

Magnetic units are defined in the following section. For conversions to and from other units representing these properties, see Chapter 2.

Magnetic flux

The standard unit of magnetic flux is the *weber* (Wb), defined as 1 V × s. This is a large unit in practice, equivalent to 1 A × H, represented by a constant, direct current of 1 A flowing through a coil having an inductance of 1 H. Magnetic flux is represented in equations by the italicized uppercase Greek letter ϕ.

Magnetic flux density

The standard unit of magnetic flux density, represented by the uppercase letter B, is the *tesla* (T), equivalent to 1 Wb/m². Sometimes, magnetic flux density is spoken of in terms of the number of *lines of flux* per unit area; this terminology is imprecise.

Magnetic field intensity

The standard unit of magnetic field intensity is the *oersted* (Oe), equivalent to 79.6 A/m. Magnetic field intensity is represented in equations by the italicized uppercase letter H.

Magnetic pole force

The standard unit of magnetic pole strength is the *ampere-meter* (A × m). Pole strength is represented in equations by the italicized lowercase p or uppercase P.

Magnetomotive force

The standard unit of magnetomotive force, symbolized by the uppercase letter F, is the *ampere-turn* (A × T), produced by a constant, direct current of 1 A flowing in a single-turn, air-core coil. Magnetomotive force is independent of coil radius.

Reluctance

The SI unit of reluctance, symbolized by the uppercase letter R, is the magnetic analog of electrical resistance. The SI unit is the *ampere-turn per Weber* (A × T/Wb). The centimeter-gram-second (cgs) unit is probably used more often; it is the *rel*, equivalent to gilberts per maxwell.

Permeability

The SI unit of permeability, symbolized by the lowercase Greek letter μ, is the extent to which a substance concentrates magnetic flux. The SI unit of permeability is the *tesla-meter per ampere* (T × m/A). The cgs unit is probably used more often; it is the *gauss per oersted* (G/Oe).

Magnetizing force

The SI unit of magnetizing force, symbolized by the uppercase letter H, is the *ampere-turn per meter* (A × T/m). Sometimes this quantity is expressed in *oersteds* (Oe).

Gain and Loss

Signal gain (amplification) and loss (attenuation) are measured in logarithmic units known as *decibels* (dB). In equations, gain is typically represented by the italicized uppercase letter G, and loss is represented by the italicized uppercase letter L.

For voltage

Let the input voltage to a circuit or system be represented by E_{in}, and the output voltage be represented by E_{out}. Assuming E_{in} and E_{out} are measured in the same units, and the input impedance is equal to the output impedance, the voltage gain in decibels, G_{VdB}, is given by:

$$G_{VdB} = 20 \times \log_{10} (E_{out}/E_{in})$$

The voltage loss in decibels, L_{VdB}, is equal to the additive inverse (negative) of G_{VdB}:

$$L_{VdB} = -G_{VdB} = 20 \times \log_{10} (E_{in}/E_{out})$$

For current

Let input current be represented by I_{in}, and output current be represented by I_{out}. Assuming I_{in} and I_{out} are measured in the same units, and the input impedance is equal to the output impedance, the current gain in decibels, G_{CdB}, is given by:

$$G_{CdB} = 20 \times \log_{10} (I_{out}/I_{in})$$

The current loss in decibels, L_{CdB}, is equal to the additive inverse (negative) of G_{CdB}:

$$L_{CdB} = -G_{CdB} = 20 \times \log_{10} (I_{in}/I_{out})$$

For power

Let input power be represented by P_{in}, and output power be represented by P_{out}. Assuming P_{in} and P_{out} are measured in the same units, the power gain in decibels, G_{PdB}, is given by:

$$G_{PdB} = 10 \times \log_{10}(P_{out}/P_{in})$$

The power loss in decibels, L_{PdB}, is equal to the additive inverse (negative) of G_{PdB}:

$$L_{PdB} = -G_{PdB} = 10 \times \log_{10}(P_{in}/P_{out})$$

Miscellaneous Units

The following units are occasionally used in electronics. For conversions to and from other units, see Chapter 2.

Area

The standard unit of area is the *square meter* or *meter squared* (m^2). Area is represented in equations by the italicized uppercase letter A.

Volume

The standard unit of volume is the *cubic meter* or *meter cubed* (m^3). Volume is represented in equations by the italicized uppercase letter V.

Plane angular measure

The standard unit of angular measure is the *radian* (rad). It is the angle subtended at the center of a circle by an arc around the perimeter of the circle, whose length, as measured on the circle, is equal to the radius of the circle. Angles are represented in equations by italicized lowercase Greek letters, usually ϕ or θ.

Solid angular measure

The standard unit of solid angular measure is the *steradian* (sr). A solid angle of 1 sr is represented by a cone with its apex at the center of a sphere, intersecting the surface of the sphere in a circle so that, within the circle, the enclosed area on the sphere is equal to the square of the radius of the sphere.

Velocity

The standard unit of linear speed is the *meter per second* (m/s). The unit of velocity requires two specifications: speed and direction. Direction is indicated in *radians* clockwise from geographic north on the earth's surface, and counterclockwise from the positive x axis in the coordinate xy-plane.

In three dimensions, direction can be specified in rectangular, spherical, or cylindrical coordinates. Speed and velocity are represented in equations by the italicized lowercase letter v.

Angular velocity

The standard unit of angular velocity is the *radian per second* (rad/s). Angular velocity is represented in equations by the italicized lowercase Greek letter ω.

Acceleration

The standard unit of acceleration is the *meter per second per second*, or *meter per second squared* (m/s^2). Linear acceleration is represented in equations by the italicized lowercase letter a.

Angular acceleration

The standard unit of angular acceleration is the *radian per second per second*, or *radian per second squared* (rad/s^2). Angular acceleration is represented in equations by the italicized lowercase Greek letter α.

Force

The standard unit of force is the *newton* (N). It is the impetus required to cause the linear acceleration of a 1-kg mass at a rate of 1 m/s^2. Force is represented in equations by the italicized uppercase letter F.

Chapter 2

Conversions and Constants

This chapter contains data for conversions between fundamental units (defined in Chapter 1) and less-common or nonstandard units for the same quantities. Physical, electrical, and chemical constants are also noted.

Prefix Multipliers

Any unit can be expressed in larger or smaller units that are multiples or fractions of the fundamental unit. These multipliers and divisors are given standard values and prefix names. *Decimal (power of 10) prefix multipliers* represent orders of magnitude in base 10 (the decimal number system); they are used in analog electronics and general science. *Binary (power of 2) prefix multipliers* represent orders of magnitude in base 2 (the binary number system); they are used in digital electronics and computer science. Table 2.1 lists prefix names and multiplication factors for both schemes.

Alternative Unit Systems

The SI System of Units is the most widely accepted system. However, other schemes are sometimes encountered. The most common among these are the *Centimeter/Gram/Second (cgs) System* and the *Foot/Pound/Second (English) System*.

SI Unit Conversions

Table 2.2 is a conversion database for SI units defined in Chapter 1, to and from various other

TABLE 2.1 Prefix Multipliers and Their Abbreviations.

Designator	Symbol	Decimal	Binary
yocto-	y	10^{-24}	2^{-80}
zepto-	z	10^{-21}	2^{-70}
atto-	a	10^{-18}	2^{-60}
femto-	f	10^{-15}	2^{-50}
pico-	p	10^{-12}	2^{-40}
nano-	n	10^{-9}	2^{-30}
micro-	µ or mm	10^{-6}	2^{-20}
milli-	m	10^{-3}	2^{-10}
centi-	c	10^{-2}	—
deci-	d	10^{-1}	—
(none)	—	10^{0}	2^{0}
deka-	da or D	10^{1}	—
hecto-	h	10^{2}	—
kilo-	K or k	10^{3}	2^{10}
mega-	M	10^{6}	2^{20}
giga-	G	10^{9}	2^{30}
tera-	T	10^{12}	2^{40}
peta-	P	10^{15}	2^{50}
exa-	E	10^{18}	2^{60}
zetta-	Z	10^{21}	2^{70}
yotta-	Y	10^{24}	2^{80}

units. The first column lists units to be converted; the second column lists units to be derived. The third column lists numbers by which units in the first column must be multiplied to obtain units in the second column. The fourth column lists num-

TABLE 2.2 SI Unit Conversions. When No Coefficient is Given, the Coefficient is Meant to Be Precisely Equal to 1.

To Convert:	To:	Multiply By:	Conversely, Multiply By:
meters (m)	Angstroms	10^{10}	10^{-10}
meters (m)	nanometers (nm)	10^{9}	10^{-9}
meters (m)	microns (µ)	10^{6}	10^{-6}
meters (m)	millimeters (mm)	10^{3}	10^{-3}
meters (m)	centimeters (cm)	10^{2}	10^{-2}
meters (m)	inches (in)	39.37	0.02540
meters (m)	feet (ft)	3.281	0.3048
meters (m)	yards (yd)	1.094	0.9144
meters (m)	kilometers (km)	10^{-3}	10^{3}
meters (m)	statute miles (mi)	6.214×10^{-4}	1.609×10^{3}
meters (m)	nautical miles	5.397×10^{-4}	1.853×10^{3}
meters (m)	light seconds	3.336×10^{-9}	2.998×10^{8}

meters (m)	astronomical units (AU)	6.685×10^{-12}	1.496×10^{11}
meters (m)	light years	1.057×10^{-16}	9.461×10^{15}
meters (m)	parsecs (pc)	3.241×10^{-17}	3.085×10^{16}
kilograms (kg)	atomic mass units (amu)	6.022×10^{26}	1.661×10^{-27}
kilograms (kg)	nanograms (ng)	10^{12}	10^{-12}
kilograms (kg)	micrograms (µg)	10^{9}	10^{-9}
kilograms (kg)	milligrams (mg)	10^{6}	10^{-6}
kilograms (kg)	grams (g)	10^{3}	10^{-3}
kilograms (kg)	ounces (oz)	35.28	0.02834
kilograms (kg)	pounds (lb)	2.205	0.4535
kilograms (kg)	English tons	1.103×10^{-3}	907.0
seconds (s)	minutes (min)	0.01667	60.00
seconds (s)	hours (h)	2.778×10^{-4}	3.600×10^{-3}
seconds (s)	days (dy)	1.157×10^{-5}	8.640×10^{4}
seconds (s)	years (yr)	3.169×10^{-8}	3.156×10^{7}
seconds (s)	centuries	3.169×10^{-10}	3.156×10^{9}
seconds (s)	millenia	3.169×10^{-11}	3.156×10^{10}

TABLE 2.2 (Continued)

To Convert:	To:	Multiply By:	Conversely, Multiply By:
degrees Kelvin (°K)	degrees Celsius (°C)	Subtract 273	Add 273
degrees Kelvin (°K)	degrees Fahrenheit (°F)	Multiply by 1.80, then subtract 459	Multiply by 0.556, then add 255
degrees Kelvin (°K)	degrees Rankine (°R)	1.80	0.556
amperes (A)	carriers per second	6.24×10^{18}	1.60×10^{-19}
amperes (A)	statamperes (statA)	2.998×10^{9}	3.336×10^{-10}
amperes (A)	nanoamperes (na)	10^{9}	10^{-9}
amperes (A)	microamperes (μA)	10^{6}	10^{-6}
amperes (A)	abamperes (abA)	0.10000	10.000
amperes (A)	milliamperes (ma)	10^{3}	10^{-3}
candela (cd)	microwatts per steradian (μW/sr)	1.464×10^{-3}	6.831×10^{-4}
candela (cd)	milliwatts per steradian (μW/sr)	1.464	0.6831

candela (cd)	lumens per steradian (lum/sr)	identical; no conversion	identical; no
candela (cd)	watts per steradian (W/sr)	1.464×10^{-3}	683.1
moles (mol)	coulombs (C)	9.65×10^{4}	1.04×10^{-5}

bers by which units in the second column must be multiplied to obtain units in the first column.

Electrical Unit Conversions

Table 2.3 is a conversion database for electrical units defined in Chapter 1, to and from various other units. The first column lists units to be converted; the second column lists units to be derived. The third column lists numbers by which units in the first column must be multiplied to obtain units in the second column. The fourth column lists numbers by which units in the second column must be multiplied to obtain units in the first column.

Magnetic Unit Conversions

Table 2.4 is a conversion database for magnetic units defined in Chapter 1, to and from various other units. The first column lists units to be converted; the second column lists units to be derived. The third column lists numbers by which units in the first column must be multiplied to obtain units in the second column. The fourth column lists numbers by which units in the second column must be multiplied to obtain units in the first column.

TABLE 2.3 Electrical Unit Conversions. When No Coefficient is Given, The Coefficient is Meant to be Precisely Equal to 1.

To Convert:	To:	Multiply By:	Conversely, Multiply By:
unit electric charges	coulombs (C)	1.60×10^{-19}	6.24×10^{18}
unit electric charges	abcoulombs (abC)	1.60×10^{-20}	6.24×10^{19}
unit electric charges	statcoulombs (stat °C)	4.80×10^{-10}	2.08×10^{9}
coulombs (C)	unit electric charges	6.24×10^{18}	1.60×10^{-19}
coulombs (C)	statcoulombs (stat C)	2.998×10^{9}	3.336×10^{-10}
coulombs (C)	abcoulombs (abC)	0.1000	10.000
joules (J)	electronvolts (eV)	6.242×10^{18}	1.602×10^{-19}
joules (J)	ergs (erg)	10^{7}	10^{-7}
joules (J)	calories (cal)	0.2389	4.1859
joules (J)	British thermal units (Btu)	9.478×10^{-4}	1.055×10^{3}
joules (J)	watt-hours (w • h)	2.778×10^{-4}	3.600×10^{3}
joules (J)	kilowatt-hours (kW • h)	2.778×10^{-7}	3.600×10^{6}

TABLE 2.3 (Continued)

To Convert:	To:	Multiply By:	Conversely, Multiply By:
volts (V)	abvolts (abV)	10^8	10^{-8}
volts (V)	microvolts (µV)	10^6	10^{-6}
volts (V)	millivolts (mV)	10^3	10^{-3}
volts (V)	stratvolts (stat V)	3.336×10^{-3}	299.8
volts (V)	kilovolts (kV)	10^{-3}	10^3
volts (V)	megavolts (MV)	10^{-6}	10^6
ohms (Ω)	abohms (abΩ)	10^9	10^{-9}
ohms (Ω)	megohms (MΩ)	10^{-6}	10^6
ohms (Ω)	kilohms (kΩ)	10^{-3}	10^3
ohms (Ω)	statohms (stat Ω)	1.113×10^{-12}	8.988×10^{11}
siemens (S)	statsiemens (stat S)	8.988×10^{11}	1.113×10^{-12}
siemens (S)	microsiemens (µS)	10^6	10^{-6}
siemens (S)	millisiemens (mS)	10^3	10^{-3}

siemens (S)	absiemens (abS)	10^{-9}	10^{9}
watts (W)	picowatts (pW)	10^{12}	10^{-12}
watts (W)	nanowatts (nW)	10^{9}	10^{-9}
watts (W)	microwatts (µW)	10^{6}	10^{-6}
watts (W)	milliwatts (mW)	10^{3}	10^{-3}
watts (W)	British thermal units per hour (Btu/hr)	3.412	0.2931
ohms (Ω)	statohms (statΩ)	1.113×10^{-12}	8.988×10^{11}
watts (W)	horsepower (hp)	1.341×10^{-3}	745.7
watts (W)	kilowatts (kW)	10^{-3}	10^{3}
watts (W)	megawatts (MW)	10^{-6}	10^{6}
watts (W)	gigawatts (GW)	10^{-9}	10^{9}
hertz (Hz)	degrees per second (deg/s)	360.0	0.002778
hertz (Hz)	radians per second (rad/s)	6.283	0.1592
hertz (Hz)	kilohertz (kHz)	10^{-3}	10^{3}
hertz (Hz)	megahertz (MHz)	10^{-6}	10^{6}
hertz (Hz)	gigahertz (GHz)	10^{-9}	10^{9}

TABLE 2.3 (Continued)

To Convert:	To:	Multiply By:	Conversely, Multiply By:
hertz (Hz)	terahertz (THz)	10^{-12}	10^{12}
farads (F)	picofarads (pF)	10^{12}	10^{-12}
farads (F)	statfarads (statF)	8.898×10^{11}	1.113×10^{-12}
farads (F)	nanofarads (nF)	10^{9}	10^{-9}
farads (F)	microfarads (µF)	10^{6}	10^{-6}
farads (F)	abfarads (abF)	10^{-9}	10^{9}
henrys (H)	nanohenrys (nH)	10^{9}	10^{-9}
henrys (H)	abhenrys (abH)	10^{9}	10^{-9}
henrys (H)	microhenrys (µH)	10^{6}	10^{-6}
henrys (H)	millihenrys (mH)	10^{3}	10^{-3}
henrys (H)	stathenrys (statH)	1.113×10^{-12}	8.898×10^{11}
volts per meter (V/m)	picovolts per meter (pV/m)	10^{12}	10^{-12}
volts per meter (V/m)	nanovolts per meter (nV/m)	10^{9}	10^{-9}

volts per meter (V/m)	microvolts per meter (µV/m)	10^6	10^{-6}
volts per meter (V/m)	millivolts per (mV/m)	10^3	10^{-3}
volts per meter (V/m)	volts per foot (v/ft)	3.281	0.3048
watts per square meter (W/m²)	picowatts per square meter (pW/m²)	10^{12}	10^{-12}
watts per square meter (W/m²)	nanowatts per square meter (pW/m²)	10^9	10^{-9}
watts per square meter (W/m²)	microwatts per square meter µW/m²)	10^6	10^{-6}
watts per square meter (W/m²)	milliwatts per square meter (mW/m²)	10^3	10^{-3}
watts per square meter (W/m²)	watts per square foot (W/ft²)	0.09294	10.76
watts per square meter (W/m²)	watts per square inch (W/in²)	6.452×10^{-4}	1.550×10^3
watts per square meter (W/m²)	watts per square centimeter (W/cm²)	10^{-4}	10^4
watts per square meter (W/m²)	watts per square millimeter	10^{-6}	10^6

TABLE 2.4 Magnetic Unit Conversions. When No Coefficient is Given, The Coefficient is Meant to be Precisely Equal to 1.

To Convert:	To:	Multiply By:	Conversely, Multiply By:
webers (Wb)	maxwells (Mx)	10^8	10^{-8}
webers (Wb)	ampere-microhenrys (A · µH)	10^6	10^{-6}
webers (Wb)	ampere-millihenrys (A · mH)	10^3	10^{-3}
webers (Wb)	unit poles	1.256×10^{-7}	7.96×10^6
teslas (T)	maxwells per square meter (Mx/m^2)	10^8	10^{-8}
teslas (T)	gauss (G)	10^4	10^{-4}
teslas (T)	maxwells per square centimeter (Mx/cm^2)	10^4	10^{-4}
teslas (T)	maxwells per square millimeter (Mx/mm^2)	10^2	10^{-2}
teslas (T)	webers per square centimeter (W/cm^2)	10^{-4}	10^4
teslas (T)	webers per square millimeter (W/mm^2)	10^{-6}	10^6
oersteds (Oe)	microampere-turns per meter (µA · T/m)	7.96×10^7	1.256×10^{-8}
oersteds (Oe)	milliampere-turns per meter (mA · T/m)	7.96×10^4	1.256×10^{-5}
oersteds (Oe)	ampere-turns per meter (A · T/m)	79.6	0.01256

TABLE 2.4 *(Continued)*

To Convert:	To:	Multiply By:	Conversely, Multiply By:
ampere-turns (A · T)	microampere-turns (µA · T)	10^6	10^{-6}
ampere-turns (A · T)	milliampere-turns (mA · T)	10^3	10^{-3}
ampere-turns (A · T)	gilberts (G)	1.256	0.796

Miscellaneous Unit Conversions

Table 2.5 is a conversion database for miscellaneous units defined in Chapter 1, to and from various other units. The first column lists units to be converted; the second column lists units to be derived. The third column lists numbers by which units in the first column must be multiplied to obtain units in the second column. The fourth column lists numbers by which units in the second column must be multiplied to obtain units in the first column.

Constants

Table 2.6 lists common physical, electrical, and chemical constants. Expressed units can be converted to other units by referring to Tables 2.2 through 2.5.

TABLE 2.5 Miscellaneous Unit Conversions. When No Coefficient is Given, The Coefficient is Meant to be Precisely Equal to 1.

To Convert:	To:	Multiply By:	Conversely, Multiply By:
square meters (m^2)	square Angstroms	10^{20}	10^{-20}
square meters (m^2)	square nanometers (nm^2)	10^{18}	10^{-18}
square meters (m^2)	square microns (μ^2)	10^{12}	10^{-12}
square meters (m^2)	square millimeters (mm^2)	10^{6}	10^{-6}
square meters (m^2)	square centimeters (cm^2)	10^{4}	10^{-4}
square meters (m^2)	square inches (in^2)	1.550×10^{3}	6.452×10^{-4}
square meters (m^2)	square feet (ft^2)	10.76	0.09294
square meters (m^2)	acres	2.471×10^{-4}	4.047×10^{3}
square meters (m^2)	hectares	10^{-4}	10^{4}
square meters (m^2)	square kilometers (km^2)	10^{-6}	10^{6}
square meters (m^2)	square statute miles (mi^{-2})	3.863×10^{-7}	2.589×10^{6}
square meters (m^2)	square nautical miles	2.910×10^{-7}	3.434×10^{6}

square meters (m²)	square light years	1.117×10^{-17}	8.951×10^{16}
square meters (m²)	square parsecs (pc²)	1.051×10^{-33}	9.517×10^{32}
cubic meters (m³)	cubic Angstroms	10^{30}	10^{-30}
cubic meters (m³)	cubic nanometers (nm³)	10^{27}	10^{-27}
cubic meters (m³)	cubic microns (µ³)	10^{18}	10^{-18}
cubic meters (m³)	cubic millimeters (mm³)	10^{9}	10^{-9}
cubic meters (m³)	cubic centimeters (cm³)	10^{6}	10^{-6}
cubic meters (m³)	milliliters (ml)	10^{6}	10^{-6}
cubic meters (m³)	liters (l)	10^{3}	10^{-3}
cubic meters (m³)	U.S. gallons (gal)	264.2	3.785×10^{-3}
cubic meters (m³)	cubic inches (in³)	6.102×10^{4}	1.639×10^{-5}
cubic meters (m³)	cubic feet (ft³)	35.32	0.02831
cubic meters (m³)	cubic kilometers (km³)	10^{-9}	10^{9}
cubic meters (m³)	cubic statute miles (mi³)	2.399×10^{-10}	4.166×10^{9}
cubic meters (m³)	cubic nautical miles	1.572×10^{-10}	6.362×10^{9}
cubic meters (m³)	cubic light seconds	3.711×10^{-26}	2.695×10^{25}

TABLE 2.5 (Continued)

To Convert:	To:	Multiply By:	Conversely, Multiply By:
cubic meters (m^3)	cubic astronomical units (AU^3)	2.987×10^{-34}	3.348×10^{33}
cubic meters (m^3)	cubic light years	1.181×10^{-48}	8.469×10^{47}
cubic meters (m^3)	cubic parsecs (pc^3)	3.406×10^{-50}	2.936×10^{49}
radians (rad)	degrees (° or deg)	57.30	0.01745
meters per second (m/s)	inches per second (in/s)	39.37	0.02540
meters per seconds (m/s)	kilometers per hour (km/hr)	3.600	0.2778
meters per second (m/s)	feet per second (ft/s)	3.281	0.3048
meters per second (m/s)	statute miles per hour (mi/hr)	2.237	0.4470

meters per second (m/s)	knots (kt)	1.942	0.5149
meters per second (m/s)	kilometers per second (km/min)	0.06000	16.67
meters per second (m/s)	kilometers per second (km/s)	10^{-3}	10^3
radians per second (rad/s)	degrees per second (°/s or deg/s)	57.30	0.01745
radians per second (rad/s)	revolutions per second (rev/s or rps)	0.1592	6.283
radians per second (rad/s)	revolutions per minute (rev/min or rpm)	$2.653 \cdot 10^{-3}$	377.0
meters per second per second (ms/^{2})	inches per second per second (in/s^2)	39.37	0.02540
meters per second (m/s^2)	feet per second per second (ft/s^2)	3.281	0.3048
radians per second (rad/s^2)	degrees per second per second (°/s$_2$ or deg/s$_2$)	57.30	0.01745

TABLE 2.5 (Continued)

To Convert:	To:	Multiply By:	Conversely, Multiply By:
radians per second per second (rad/s^2)	revolutions per second per second (rv/s^2 or rps/s)	0.1592	6.283
radians per second per second (rad/s^2)	revolutions per minute per second (rev/min/s or rpm/s)	$2.653 \cdot 10^{-3}$	377.0
newtons (N)	dynes	10^5	10^{-5}
newtons (N)	ounces (oz)	3.597	0.2780
newtons (N)	pounds (lb)	0.2248	4.448

Conversions and Constants

TABLE 2.6 Physical, Electrical, and Chemical Constants.

Quantity or Phenomenon	Value	Symbol
Mass of sun	1.989×10^{30} kg	m_{sun}
Mass of earth	5.974×10^{24} kg	m_{earth}
Avogadro's number	6.022169×10^{23}	N
Mass of moon	7.348×10^{22} kg	m_{moon}
Mean radius of sun	6.970×10^{8} m	r_{sun}
Speed of electromagnetic-field propagation in free space	2.99792×10^{8} m/s	c
Faraday constant	9.649×10^{7} C/kmol	F
Mean radius of earth	6.371×10^{6} m	r_{earth}
Mean orbital speed of earth	2.978×10^{4} m/s	
Base of natural logarithms	2.718282	e or ε
Mean radius of moon	1.738×10^{6} m	r_{moon}
Characteristic impedance of free space	376.7 Ω	Z_0
Speed of sound in dry air at standard atmospheric temperature and pressure	344 m/s	
Gravitational acceleration at sea level	9.8067 m/s^2	g
Gas constant	8.3145 J/°K $\times$ mol	R_0
Wien's constant	0.0029 m $\times$ °K	σ_w
Second radiation constant	0.01439 m $\times$ °K	c_2
Permeability of free space	1.257×10^{-6} H/m	μ_0

TABLE 2.6 *(Continued)*

Quantity or Phenomenon	Value	Symbol
Stephan-Boltzmann constant	5.6697×10^{-8} W/m^2/°K	σ
Gravitational constant	6.673×10^{-11} N · m^2/kg^2	G
Permittivity of free space	8.85×10^{-12} F/m	ε_0
Boltzmann's constant	1.3807×10^{-23} J/°K	k
First radiation constant	4.993×10^{-24} J · m	c_1
Mass of alpha particle at rest	6.64×10^{-27} kg	m_α
Mass of neutron at rest	1.675×10^{-27} kg	m_n
Mass of proton at rest	1.673×10^{-27} kg	m_p
Mass of electron at rest	9.109×10^{-31} kg	m_e
Planck's constant	6.6261×10^{-34} J · s	h

Chapter 3

Mathematical Notation

This chapter denotes mathematical symbols used in electronics, along with quantities, variables, and phenomena that they commonly represent.

Greek Alphabet

Table 3.1 lists uppercase Greek letters, character names (as written in English), and usages. Table 3.2 lists lowercase Greek letters, character names, and usages. Uppercase or lowercase Greek letters

TABLE 3.1 Uppercase Greek Alphabet

Symbol	Character name	Common representations
A	alpha	—
B	beta	magnetic flux density
Γ	gamma	gamma match; general index set; curve; contour
Δ	delta	delta match; three-phase AC circuit with no common ground; increment; difference sequence; Laplacian
E	epsilon	voltage; energy
Z	zeta	impedance
H	eta	efficiency
Θ	theta	order
I	iota	current
K	kappa	magnetic susceptibility; degrees Kelvin
Λ	lambda	general index set
M	mu	mutual inductance
N	nu	Avogadro's number (6.022169×10^{23})
Ξ	xi	—
O	omicron	order
Π	pi	product; infinite product; homotopy
P	rho	Power
Σ	sigma	summation; series; infinite series
T	tau	time constant; temperature
Y	upsilon	—

Mathematical Notation

TABLE 3.1 (*Continued*)

Symbol	Character name	Common representations
Φ	phi	magnetic flux; Frattini subgroup
Χ	chi	reactance
Ψ	psi	dielectric flux
Ω	omega	ohms; volume of a body

TABLE 3.2 Lowercase Greek Alphabet.

Symbol	Character name	Common representations
α	alpha	current gain of bipolar transistor in common-base configuration; alpha particle; angular acceleration; angle; direction angle; transcendental number; scalar coefficient
β	beta	current gain of bipolar transistor in common-emitter configuration; magnetic flux density; beta particle; angle; direction angle; transcendental number; scalar coefficient
γ	gamma	gamma radiation; electrical conductivity; Euler's constant; gravity; direction angle; scalar coefficient; permutation; cycle

TABLE 3.2 (*Continued*)

Symbol	Character name	Common representations
δ	delta	derivative; variation of a quantity; point evaluation; support function; metric function; distance function; variation of an integral; Laplacian
ε	epsilon	electric permittivity; natural logarithm base (approximately 2.71828); eccentricity; signature
ζ	zeta	impedance, coefficient; coordinate variable in a transformation
η	eta	electric susceptibility; hysteresis coefficient; efficiency; coordinate variable in a transformation
θ	theta	angle; phase angle; angle in polar coordinates; angle in cylindrical coordinates; angle in spherical coordinates; parameter; homomorphism
ι	iota	definite description (in predicate logic)
κ	kappa	dielectric constant; coefficient of coupling; curvature
λ	lambda	wavelength; Wien Displacement Law constant; ratio; Lebesgue measure; eigenvalue
μ	mu	micro-; magnetic permeability; amplification factor; charge carrier mobility; mean; statistical parameter

Mathematical Notation

TABLE 3.2 (*Continued*)

Symbol	Character name	Common representations
υ	nu	frequency; reluctivity; statistical parameter; natural epimorphism
ξ	xi	coordinate variable in a transformation
o	omicron	order
π	pi	ratio of circle circumference to diameter (approximately 3.14159); radian; permutation
ρ	rho	electrical resistivity; variable representing an angle; curvature; correlation; metric; density
σ	sigma	electrical conductivity; Stefan-Boltzmann constant; standard deviation; variance; mathematical partition; permutation; topology
τ	tau	time-phase displacement; torsion; mathematical partition; topology
υ	upsilon	—
ϕ or φ	phi	angle; phase angle; dielectric flux; angle in spherical coordinates; Euler phi function; mapping; predicate
χ	chi	magnetic susceptibility; characteristic function; chromatic number; configuration of a body
ψ	psi	angle; mapping; predicate; chart
ω	omega	angular velocity; period; modulus of continuity

Chapter Three

are sometimes italicized. In these tables, characters are not italicized.

General Symbols

Table 3.3 lists symbols used to depict operations, relations, and specifications in mathematics relevant to physical sciences and engineering. Italics are sometimes used for alphabetic characters; other symbols are rarely italicized.

TABLE 3.3 General Mathematical Symbols and Their Common Meanings. For Meanings of Greek Letters, Refer to Tables 3.1 and 3.2.

Symbol	Character name	Common representations
.	decimal or radix point	separates integral part of number from fractional part
$\forall$	universal qualifier	read "for all"
#	pound sign	number; pounds
$\exists$	existential qualifier	read "there exists" or "for some"
%	per cent sign	read "parts per hundred" or "percent"
‰	per mil sign	read "parts per thousand" or "permil"

Mathematical Notation

TABLE 3.3 (*Continued*)

Symbol	Character name	Common representations
&	ampersand	logical AND operation
@	at sign	read "at the rate of" or "at the cost of"
( )	parentheses	encloses elements defining coordinates of a point; encloses elements of a set of ordered numbers; encloses bounds of an open interval
[ ]	brackets	encloses a group of terms that includes one or more groups in parentheses; encloses elements of an equivalence class; encloses bounds of a closed interval
{ }	braces	encloses a group of terms that includes one or more groups in brackets; encloses elements comprising a set
[ ) or ( ]	half-brackets	encloses bounds of a half-open interval
] [	inside-out brackets	encloses bounds of an open interval
$\begin{pmatrix} \end{pmatrix}$ or $\begin{bmatrix} \end{bmatrix}$	parenthesis or brackets (enlarged)	encloses elements of a matrix
*	asterisk	multiplication; logical AND operation
×	cross	multiplication; logical AND operation; vector (cross) product of two vectors

TABLE 3.3 (*Continued*)

Symbol	Character name	Common representations
Π	uppercase Greek letter pi (enlarged)	product of many values
·	small dot	multiplication
•	large dot	logical AND operation; scalar (dot) product of two vectors
+	plus sign	addition; logical OR operation
Σ	uppercase Greek letter sigma (enlarged)	summation of many values
,	comma	separates large numbers by thousands; separates elements defining coordinates of a point; separates elements of a set of ordered numbers; separates bounds of an interval
−	minus sign	subtraction; logical NOT symbol
±	plus/minus sign	read "plus or minus" and defines the extent to which a value can deviate from the nominal value
/	slash or slant	division; ratio; proportion; separates parts of a Web site uniform resource locator (URL)

Mathematical Notation

TABLE 3.3 (*Continued*)

Symbol	Character name	Common representations
$\div$	—	division
:	colon	ratio; separates hours from minutes; separates minutes from seconds
::	double colon	mean
!	exclamation mark	factorial
$\leq$	inequality sign	read "is less than or equal to"
$<$	inequality sign	read "is less than"
$<<$	inequality sign	read "is much less than"
$=$	equal sign	read "is equal to"; logical equivalence
$\geq$	inequality sign	read "is greater than or equal to"
$>$	inequality sign	read "is greater than"
$>>$	inequality sign	read "is much greater than"
$\cong$	congruence sign	read "is congruent with"
$\neq$	unequal sign	read "is not equal to"
$\equiv$	equivalence sign	read "is logically equivalent to"

TABLE 3.3 *(Continued)*

Symbol	Character name	Common representations
≈	approximation sign	read "is approximately equal to"
∝	—	read "is proportional to"
~	squiggle	read "is similar to"
...	triple dot	read "and so on" or "and beyond"
|	vertical line	read "is exactly divisible by"
| |	vertical lines	absolute value of quantity between lines; length of vector quantity denoted between lines; distance between two points; cardinality of number; modulus
|	vertical line (elongated)	denotes limits of evaluation for a function
||	vertical lines (elongated)	determinant of matrix whose elements are enumerated between lines
ℵ	uppercase Hebrew letter aleph	transfinite cardinal number; Continuum Hypothesis
∩	intersection sign	set-intersection operation
∪	union sign	set-union operation
∅	null sign	set containing no elements (empty set or null set)

Mathematical Notation 53

TABLE 3.3 *(Continued)*

Symbol	Character name	Common representations
$\in$	—	read "is an element of"
$\notin$	—	read "is not an element of"
$\subset$	—	read "is a proper subset of"
$\supset$	implication sign	read "logically implies"
$\subseteq$	—	read "is a subset of"
$\not\subset$	—	read "is not a proper subset of"
$\angle$	angle sign	angle; angle measure
$\perp$	—	read "is perpendicular to"
∇	del or nabla	vector differential operator
$\sqrt{}$	radical or surd	root; square root
$\Leftrightarrow$ or $\leftrightarrow$	double-arrow	read "if and only if" or "is logically equivalent to"
$\Rightarrow$	right arrow	logical implication
$\therefore$	three dots	read "therefore"
$\rightarrow$	right arrow	logical implication; convergence; mapping
$\uparrow$	upward arrow	read "above" or "increasing"
$\downarrow$	downward arrow	read "below" or "decreasing"
∂	—	partial derivative; Jacobian; surface of a body

TABLE 3.3 (Continued)

Symbol	Character Name	Common Representations
$\int$	—	integral
$\iint$	—	double integral
$\int_E$	—	Riemann integral
$\int_\Gamma$	—	contour integral
$\iint_S$	—	surface integral
$\iiint$	—	triple integral
°	degree sign (superscript)	degree of angle; degree of temperature
∞	infinity sign	infinity; an arbitrarily large number; an arbitrarily great distance away

Subscripts and Superscripts

Subscripts modify the meanings of units, constants, and variables. A subscript is placed to the right of the main character (without spacing), is set in smaller type than the main character, and is set below the base line. Numeric subscripts are generally not italicized; alphabetic subscripts are sometimes italicized. Examples of subscripted quantities are:

Z_0 read "Z sub nought"; represents characteristic impedance

R_{out} read "R sub out"; stands for output resistance

x_3 read "x sub 3"; represents a variable

Superscripts represent exponents (the taking of the base quantity or variable to the indicated power). Superscripts are usually numerals, but they are sometimes alphabetic characters. Italicized, lowercase English letters from the second half of the alphabet (n through z) are generally used to represent variable exponents. A superscript, placed to the right of the main character (without spacing), is set in smaller type than the main character, and is set above the base line. Examples of superscripted quantities are:

2^3 read "two cubed"; represents $2 \times 2 \times 2$

e^x read "e to the xth"; represents the exponential function of x

$y^{1/2}$ read "y to the one half"; represents the square root of y

Scientific Notation

Scientific notation is used to represent extreme numerical values. It also facilitates arithmetic operations among numbers ranging over many orders of magnitude. A numeral in scientific notation is written in the form:

$$m.n \times 10^z$$

where m (to the left of the radix point) is a number from the set {1, 2, 3, 4, 5, 6, 7, 8, 9}, n (to the right of the radix point) is a non-negative integer, and z (the power of 10) can be any integer. Some examples of numbers written in scientific notation are:

$$2.56 \times 10^6$$
$$8.0773 \times 10^{-18}$$
$$1.000 \times 10^0$$

In some countries, scientific notation requires that $m = 0$. In this rarely used form, the previous numbers appear as:

$$0.256 \times 10^7$$
$$0.80773 \times 10^{-17}$$
$$0.1000 \times 10^1$$

The multiplication sign can be denoted in various ways. Instead of the times sign (×), an asterisk (*) can be used, so these expressions become:

$$2.56 * 10^6$$
$$8.0773 * 10^{-18}$$
$$1.000 * 10^0$$

Another alternative is to use the dot (·), so the expressions appear as:

$$2.56 \cdot 10^6$$
$$8.0773 \cdot 10^{-18}$$
$$1.000 \cdot 10^0$$

Sometimes it is necessary to express numbers in scientific notation using plain text. This is the case, for example, when transmitting information within the body of an e-mail message (rather than as an attachment). Some electronic calculators and computers use this system. The uppercase letter E indicates that the quantity immediately following is an exponent. In this format, the previous expressions are written as:

$$2.56\,\text{E}6$$
$$8.0773\,\text{E}{-18}$$
$$1.000\,\text{E}0$$

58   Chapter Three

Sometimes the exponent is always written with two numerals and always includes a plus sign or a minus sign, so the previous expressions appear as:

$$2.56\text{E}+06$$
$$8.0773\text{E}-18$$
$$1.000\text{E}+00$$

Another alternative is to use an asterisk to indicate multiplication, and the symbol ^ to indicate a superscript, so the expressions appear as:

$$2.56 * 10\text{^}6$$
$$8.0773 * 10\text{^}-18$$
$$1.000 * 10\text{^}0$$

In all of these examples, the numerical values represented are identical. Respectively, if written out in full, they are:

$$2{,}560{,}000$$
$$0.0000000000000000080773$$
$$1.000$$

In printed literature, it is common to use scientific notation only when z (the power of 10) is fairly large or small. If $-2 \leq z \leq 2$, numbers are written out in full as a rule, and the power of 10 is not shown. If $z = -3$ or $z = 3$, numbers are sometimes written out in full and are sometimes depicted in scientific notation. If $z \leq -4$ or $z \geq 4$, values are expressed in scientific notation as a rule. Calculators set to display quantities in scientific notation will usually show the power of 10 for all numbers, even those for which the power of 10 is zero.

Addition and subtraction of numbers is best done by writing numbers out in full, if possible. Thus, for example:

$$(3.045 \times 10^2) + (6.853 \times 10^3)$$
$$= 304.5 + 6853 = 7157.5$$
$$= 7.1575 \times 10^3$$

When numbers are multiplied or divided in scientific notation, the decimal numbers (to the left of the multiplication symbol) are multiplied or divided by each other. Then the powers of 10 are added (for multiplication) or subtracted (for division). Finally, the product or quotient is reduced to standard form. An example is:

$$(3.045 \times 10^2)\,(6.853 \times 10^3)$$
$$= 20.867385 \times 10^5 = 2.0867385 \times 10^6$$

Significant Figures

In scientific notation, the term *significant figures* refers to the number of numerals in the decimal portion of an expression that can be relied upon to portray the quantity to a known degree of accuracy. For example, 3.83×10^{-25} has three significant digits, while 3.83018×10^{-25} has six significant digits and portrays the quantity to a greater degree of accuracy.

Truncation

The process of *truncation* deletes all the numerals to the right of a certain point in the decimal part of an expression. Many, if not most, electronic calculators use this process to fit numbers within their displays. For example, the number 3.830175692803 can be shortened in steps as follows:

$$3.830175692803$$
$$3.83017569280$$
$$3.8301756928$$
$$3.830175692$$
$$3.83017569$$
$$3.8301756$$
$$3.830175$$
$$3.83017$$

3.83
3.8
3

Rounding

Rounding is a more accurate, and preferred, method of rendering numbers in shortened form. In this process, when a given digit (call it r) is deleted at the right-hand extreme of an expression, the digit (q) to its left (which becomes the new r after the old r is deleted) is not changed if $0 \leq r \leq 4$. If $5 \leq r \leq 9$, then q is increased by 1 ("rounded up"). Some electronic calculators use rounding rather than truncation. If rounding is used, the number 3.830175692803 can be shortened in steps as follows:

3.830175692803
3.83017569280
3.8301756928
3.830175693
3.83017569
3.8301757
3.830176
3.83018
3.8302

3.830
3.83
3.8
4

In calculations

When calculations are performed using scientific notation, the number of significant figures in the result cannot be greater than the number of significant figures in the shortest expression in the calculation.

In the foregoing example showing addition, the sum, 7.1575×10^3, must be cut down to four significant figures because the addends have only four significant figures. If the resultant is truncated, it becomes 7.157×10^3. If rounded, it becomes 7.158×10^3.

In the foregoing example showing multiplication, the resultant, 2.0867385×10^6, must be cut down to four significant figures because the multiplicands have only four significant figures. If the resultant is truncated, it becomes 2.086×10^6. If rounded, it becomes 2.087×10^6.

"Downsizing" of resultants is best done at the termination of a calculation process, if that process involves more than one computation.

Precedence of Operations

Unless otherwise indicated by parentheses and/or brackets, equations and formulas containing various operations are resolved by performing operations in the following order:

- Exponentiation and functions (e.g., x^3, $\sin x$, or e^x)
- Multiplication and division (e.g., $x \times y$ or $x/3$)
- Addition and subtraction (e.g., $x + y$ or $x - 3$)

Thus, for example, the following two expressions are equivalent:

$$3 \times x^3 - 4 \times x^2 + \sin x + 3/x - 7$$
$$[3 \times (x^3)] - [4 \times (x^2)] + (\sin x) + (3/x) - 7$$

The parentheses and brackets in the second expression are technically not necessary. However, parentheses and brackets are often used for clarity even when they are not strictly required. They must be used when the operations in an expression are performed in an order other than would be the case if no parenthesis or brackets were used. Thus, for example, the parentheses and brackets are essential in the following expression:

$$3 \times (x^3 - 4) \times x^2 + \sin (x + 3)/(x - 7)$$

Chapter 4

Algebra and Trigonometry

This chapter contains information about basic algebra, coordinate systems, trigonometric functions and formulas, and logarithmic functions and formulas.

Theorems in Algebra

Some common *theorems*, also called *rules* or *laws*, in algebra are depicted in Table 4.1. These theo-

Chapter Four

TABLE 4.1 Common Theorems in Algebra. These Theorems Apply to All Real Numbers As Long As the Denominator is Nonzero.

Equation	Description
$x + 0 = x$	additive identity
$x \times 1 = x$	multiplicative identity
$x \times 0 = 0$	multiplication by zero
$-(-x) = x$	double negation
$x + (-x) = 0$	additive inverse
$x \times (1/x) = 1$	multiplicative inverse
$1 \times (1/x) = x$	reciprocal-of-a-reciprocal rule
$x + y = y + x$	commutative law of addition
$x \times y = y \times x$	commutative law of multiplication
$x + (y + z) = (x + y) + z$	associative law of addition
$x \times (y \times z) = (x \times y) \times z$	associative law of multiplication
$x \times (y + z) = x \times y + x \times z$	distributive law
$(w + x) \times (y + z) =$ $w \times y + w \times z +$ $x \times y + x \times z$	product of sums
$w/x = y/z \rightarrow w \times z = x \times y$	cross-multiplication rule
$1/(x \times y) = (1/x) \times (1/y)$	reciprocal of a product
$1/(x/y) = y/x$	reciprocal of a quotient

rems apply to all real numbers, with one exception: when a variable appears as the denominator of a quotient (for example, $1/x$), the expression is undefined for $x = 0$.

Coordinate Systems

Relations and functions are commonly plotted in *coordinate systems*. These schemes show characteristics of devices and phenomena, such as antenna radiation patterns, waveforms, and spectral displays.

Cartesian plane

The most common two-dimensional coordinate system is the *Cartesian plane* (Figure 4.1), also called *rectangular coordinates* or the *xy-plane*. The independent variable is plotted along the x axis or *abscissa*; the dependent variable is plotted along the y axis or *ordinate*. The scales of the abscissa and ordinate are normally linear, although the divisions need not represent the same increments. Variations of this scheme include the *semilog graph*, in which the ordinate scale is logarithmic, and the *log-log graph*, in which both scales are logarithmic.

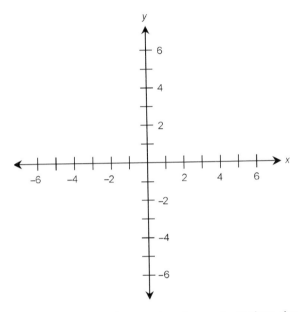

Figure 4.1 The Cartesian or rectangular coordinate plane, also called the *xy-plane*.

Polar coordinate plane

Another two-dimensional system is the *polar coordinate plane*. The independent variable is plotted as the radius (r) and the dependent variable is plotted as an angle (θ). Figure 4.2A shows the polar system used in mathematics and physical sciences; θ is

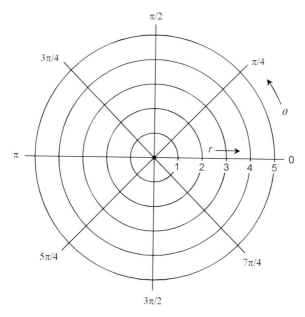

Figure 4.2A The polar plane for mathematics and physical sciences.

in radians and is plotted counterclockwise from the ray extending to the right ("east"). Figure 4.2B shows the polar system used in wireless communications, navigation, and location applications; θ is in degrees and is plotted clockwise from the ray extending upwards ("north"). The angular scale is al-

70   Chapter Four

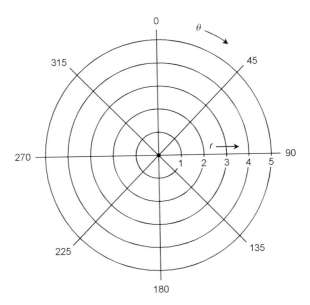

Figure 4.2B The polar plane for wireless communications, location, and navigation.

ways linear in any polar system. The radial scale is linear in most polar graphs, but, in some cases, it is logarithmic.

Latitude and longitude

Latitude and *longitude* angles uniquely define the positions of points on the surface of a sphere or in the sky. The scheme for geographic locations on the earth is illustrated in Figure 4.3A. The *polar axis* connects two specified points at antipodes on the sphere. These points are assigned latitude $\theta = 90°$ (north pole) and $\theta = -90°$ (south pole). The *equatorial axis* runs outward from the center of the sphere at a 90° angle to the polar axis. It is assigned longitude $\phi = 0°$. Latitude θ is measured positively (north) and negatively (south) relative to the plane of the equator. Longitude (ϕ) is measured counterclockwise (east) and clockwise (west) relative

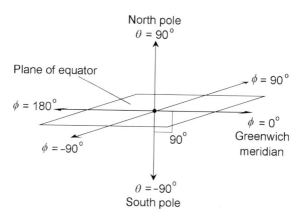

Figure 4.3A Scheme for latitude and longitude.

to the equatorial axis. The angles are restricted as follows:

$$-90° \leq \theta \leq 90°$$
$$-180° \leq \phi \leq 180°$$

On the earth's surface, the half-circle connecting the 0° longitude line with the poles passes through Greenwich, England and is known as the *Greenwich meridian*. Longitude angles are defined with respect to this meridian.

Celestial coordinates

Celestial latitude and *celestial longitude* are extensions of the earth's latitude and longitude into the heavens. Figure 4.3A applies to this system. An object whose celestial latitude and longitude coordinates are (θ,ϕ) appears at the zenith in the sky from the point on the earth's surface whose latitude and longitude coordinates are (θ,ϕ).

Declination and *right ascension* define the positions of objects in the sky relative to the stars. Figure 4.3B applies to this system. Declination (θ) is identical to celestial latitude. Right ascension (ϕ) is measured eastward from the *vernal equinox* (the position of the sun in the heavens at the moment spring begins in the northern hemisphere). The angles are restricted as follows:

Algebra and Trigonometry 73

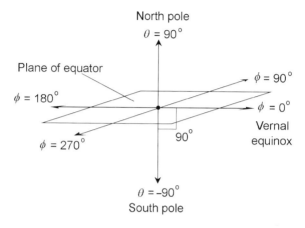

Figure 4.3B Scheme for declination and right ascension.

$$-90° \leq \theta \leq 90°$$
$$0° \leq \theta < 360°$$

Cartesian three space

An extension of rectangular coordinates into three dimensions is *Cartesian three-space* (Figure 4.4), also called *xyz-space*. Independent variables are usually plotted along the x and y axes; the dependent variable is plotted along the z axis. The scales are normally linear, although the

Chapter Four

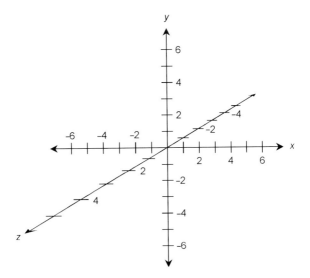

Figure 4.4 Cartesian three-space, also called *xyz-space*.

divisions need not represent the same increments. Variations of this scheme can use logarithmic graduations for one, two, or all three scales.

Cylindrical coordinates

Figure 4.5 shows a system of *cylindrical coordinates* for specifying the positions of points in

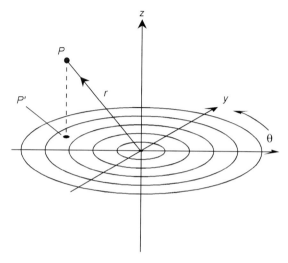

Figure 4.5 Cylindrical coordinates for defining points in three-space.

three-space. Given a set of Cartesian coordinates or xyz-space, an angle (θ) is defined in the xy-plane, measured in radians counterclockwise from the x axis. Given a point (P) in space, consider its projection (P') onto the xy-plane. The position of P is defined by the ordered triple (θ, r, z) such that:

Chapter Four

θ = Angle between P' and the x axis in the xy-plane

r = Distance (radius) from P to the origin

z = Distance (altitude) of P above the xy-plane

Spherical coordinates

Figure 4.6 shows a system of *spherical coordinates* for defining points in space. This scheme is identical to the system for declination and right

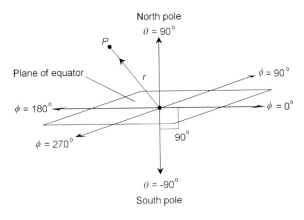

Figure 4.6 Spherical coordinates for defining points in three-space.

ascension, with the addition of a radius vector (r) representing the distance of point P from the origin. The location of point P is defined by the ordered triple (θ,ϕ,r) such that:

θ = declination of P

ϕ = right ascension of P

r = distance (radius) from P to the origin

In this example, angles are specified in degrees; alternatively, they can be expressed in radians. The several variations of this system are all commonly called *spherical coordinates*.

Trigonometry

The three basic *trigonometric functions* are: the *sine* function, the *cosine* function, and the *tangent* function. These functions typically apply to angles θ in radians so that $0 \leq \theta < 2\pi$. In formulas and equations, these functions are abbreviated *sin θ*, *cos θ*, and *tan θ*.

Basic functions

Consider a circle in the Cartesian plane with the following equation:

$$x^2 + y^2 = 1$$

This is called the *unit circle* because its radius is one unit and it is centered at the origin (0,0). Let θ be an angle whose apex is at the origin and that is measured counterclockwise from the abscissa (x axis). Suppose that this angle corresponds to a ray that intersects the unit circle at some point $P = (x_0, y_0)$. Then:

$$y_0 = \sin \theta$$
$$x_0 = \cos \theta$$
$$y_0/x_0 = \tan \theta$$

Secondary functions

Three more trigonometric functions are derived from those defined in the last section. They are the *cosecant* function, the *secant* function, and the *cotangent* function. In formulas and equations, these functions are abbreviated *csc* θ, *sec* θ, and *cot* θ. They are defined as:

$$\csc \theta = 1/(\sin \theta) = 1/y_0$$
$$\sec \theta = 1/(\cos \theta) = 1/x_0$$
$$\cot \theta = 1/(\tan \theta) = x_0/y_0$$

Trigonometric identities

Theorems in trigonometry, known as *trigonometric identities*, have been demonstrated. Some common identities are listed in Table 4.2. All angles (θ) are in radians. Note the following standard abbreviations:

$$\sin^2 \theta = (\sin \theta)^2 = (\sin \theta) \times (\sin \theta)$$
$$\cos^2 \theta = (\cos \theta)^2 = (\cos \theta) \times (\cos \theta)$$
$$\tan^2 \theta = (\tan \theta)^2 = (\tan \theta) \times (\tan \theta)$$
$$\csc^2 \theta = (\csc \theta)^2 = (\csc \theta) \times (\csc \theta)$$
$$\sec^2 \theta = (\sec \theta)^2 = (\sec \theta) \times (\sec \theta)$$
$$\cot^2 \theta = (\cot \theta)^2 = (\cot \theta) \times (\cot \theta)$$

Logarithms

A *logarithm* is an exponent to which a constant is raised to obtain a given number. Suppose that the following relationship exists among three real numbers a, m, and x:

$$a^m = x$$

Then m is the logarithm of x in base a. This expression is written:

$$m = \log_a x$$

Chapter Four

TABLE 4.2 Common Trigonometric Identities. These Theorems Apply to All Real Numbers As Long As the Denominator is Nonzero.

Equation	Description
$\sin^2 \theta + \cos^2 \theta = 1$	theorem of Pythagoras for sine and cosine functions
$\sec^2 \theta - \tan^2 \theta = 1$	theorem of Pythagoras for secant and tangent functions
$\sin - \theta = -\sin \theta$	sines of negative angles
$\cos - \theta = \cos \theta$	cosines of negative angles
$\tan - \theta = -\tan \theta$	tangents of negative angles
$\sin (\theta + 2 \times \pi) - \sin \theta$	periodicity of sine function
$\cos (\theta + 2 \times \pi) = \cos \theta$	periodicity of cosine function
$\tan (\theta + 2 \times \pi) = \tan \theta$	periodicity of tangent function
$\sin (2 \times \theta) = 2 \times \sin \theta \times \cos \theta$	double-angle formula for sine function
$\cos (2 \times \theta)$ $= 1 - (2 \times \sin^2 \theta)$ $= (2 \times \cos^2 \theta) - 1$	double-angle formula for cosine function
$\tan (2 \times \theta) = (2 \times \tan \theta)$ $(1 - \tan^2 \theta)$	double-angle formula for tangent function
$\sin (\theta/2) = \pm [(1 - \cos \theta)/2]^{1/2}$	half-angle formula for sine function
$\cos (\theta/2) = \pm [(1 + \cos \theta)/2]^{1/2}$	half-angle formula for cosine function
$\tan (\theta/2) = (\sin \theta) / (1 + \cos \theta)$	half-angle formula for tangent function

Algebra and Trigonometry 81

TABLE 4.2 (Continued)

Equation	Description
$\sin(\theta + \phi) = \sin\theta \times \cos\phi + \cos\theta \times \sin\phi$	sum formula for sine function
$\cos(\theta + \phi) = \cos\theta \times \cos\theta - \sin\theta \times \sin\phi$	sum formula for cosine function
$\tan(\theta + \phi) = (\tan\theta + \tan\phi) / (1 - \tan\theta \times \tan\phi)$	sum formula for tangent function
$\sin(\theta - \phi) = \sin\theta \times \cos\phi - \cos\theta \times \sin\phi$	difference formula for sine function
$\cos(\theta - \phi) = \cos\theta \times \cos\theta + \sin\theta \times \sin\phi$	difference formula for cosine function
$\tan(\theta - \phi) = (\tan\theta - \tan\phi) / (1 + \tan\theta \times \tan\phi)$	difference formula for tangent function
$\sin\theta = \cos(\pi/2 - \theta)$	complementary-angle rule for sine and cosine functions
$\cos\theta = \sin(\pi/2 - \theta)$	complementary-angle rule for sine and cosine functions
$\tan\theta = \cot(\pi/2 - \theta)$	complementary-angle rule for tangent and cotangent functions
$\cot\theta = \tan(\pi/2 - \theta)$	complementary-angle rule for tangent and cotangent functions

82 Chapter Four

TABLE 4.2 (Continued)

Equation	Description
sec θ = csc (π/2 − θ)	complementary-angle rule for secant and cosecant functions
csc θ = sec (π/2 − θ)	complementary-angle rule for secant and cosecant functions
sin θ = sin (π − θ)	supplementary-angle rule for sine function
cos θ = −cos (π − θ)	supplementary-angle rule for cosine function
tan θ = tan (π − θ)	supplementary-angle rule for tangent function

The two most common logarithm bases are $a = 10$ and $a = e \approx 2.71828$.

Base-10 logarithms are also called *common logarithms*. In equations, common logarithms are denoted by writing *log*. For example:

$$\log 10 = 1.000$$

Base-*e* logarithms are also known as *natural logarithms* or *Napierian logarithms*. In equations, the natural-logarithm function is usually abbreviated ln, although it is sometimes denoted as log_e. Hence, for example:

$$\ln 2.71828 = \log_e 2.71828 \approx 1.00000$$

Table 4.3 lists some formulas and relations involving logarithms.

TABLE 4.3 Common Logarithmic Identities. These Theorems Apply to All Real Numbers As Long As the Denominator is Nonzero.

Equation	Description
$\log x = \ln x / \ln 10 \approx 0.434 \times \ln x$	conversion of natural (base-e) logarithm to common (base-10) logarithm
$\ln x = \log x / \log e \approx 2.303 \times \log x$	conversion of common logarithm to natural logarithm
$\log (x \times y) = \log x + \log y$	common logarithm of a product
$\ln (x \times y) = \ln x + \ln y$	natural logarithm of a product
$\log (x/y) = -\log (y/x) = \log x - \log y$	common logarithm of a quotient
$\ln (x/y) = -\ln (y/x) = \ln x - \ln y$	natural logarithm of a quotient
$\log x^y = y \times \log x$	common logarithm of a power
$\ln x^y = y \times \ln x$	natural logarithm of a power

TABLE 4.3 (*Continued*)

Equation	Description
$\log 1/x = -\log x$	common logarithm of a reciprocal
$\ln 1/x = -\ln x$	natural logarithm of a reciprocal
$\log (x)^{1/y} = (\log x)/y$	common logarithm of a root
$\ln (x)^{1/y} = (\ln x)/y$	natural logarithm of a root
$\log 10^x = x$	common logarithm of base-10 exponential function
$\ln e^x = x$	natural logarithm of base-e exponential function

Chapter 5

Sequences and Series

This chapter contains definitions and formulas for sequences, progressions, and series. At the end of the chapter, several formulas, which are derived from series expansions of exponential and trigonometric functions, are listed.

Definitions

Some important definitions follow. The number of terms (numbers or variables) is infinite unless otherwise indicated.

Sequence

A *sequence* consists of terms in a specific order:

$$a_1, a_2, a_3, ..., a_n, ...$$

Occasionally, the expression sequence, used without adjectives, refers to the positive integers:

$$a_1, a_2, a_3, ..., a_n, ... = 1, 2, 3, ..., n, ...$$

Usually, the numbers in a sequence proceed according to a more subtle pattern.

Progression

A *progression* is a sequence in which each term is derived from the previous term, according to a formula. If $a_{(n+1)} = f(a_n)$, where f is the formula that determines each subsequent term, then the progression P, written in full, is:

$$P = a_1, f(a_1), f[f(a_1)], f\{f(f(a_1))\}, ...$$

Series

A *series* consists of numbers or variables arranged in a specific order, separated by addition symbols:

$$a_1 + a_2 + a_3 + ... + a_n + ...$$

In most applications, the terms in a series proceed according to a pattern.

Partial sum

The *partial sum* S_n of a series is the sum of its first n terms:

$$S_n = a_1 + a_2 + a_3 + ... + a_n$$

Convergent series

A *convergent series* is a series whose partial sum S_n approaches a specific finite number S as n increases without bound:

$$S_n \to S \text{ as } n \to \infty$$
$$\therefore$$
$$S = a_1 + a_2 + a_3 + ...$$

An example of a convergent series is:

$$C = 1 + 1/2 + 1/4 + 1/8 + ... = 2$$

Divergent series

A *divergent series* is a series that is not convergent for any parameter; its partial sum S_n does not approach any specific finite number as n increases without bound. An example of a divergent series is:

$$D = 1 + 2 + 3 + 4 + ...$$

Conditionally convergent series

A *conditionally convergent series* is a series that is convergent for certain values of a parameter x, but is divergent for other values of x. An example of a conditionally convergent series is:

$$C_C = 1 - x + x^2 - x^3 + x^4 - x^5 + ...$$

Converges to $1/(1+x)$ if $-1 < x < 1$

Diverges if $x \leq -1$ or $x \geq 1$

Factorial

For a given positive integer (n), the number n *factorial* (written $n!$) is the product of all the positive integers up to, and including, n:

$$n! = 1 \times 2 \times 3 \times 4 \times ... \times n$$

Factorials of n become large rapidly as n increases. Factorials of the first several positive integers are:

$$1! = 1$$
$$2! = 2$$
$$3! = 6$$
$$4! = 24$$
$$5! = 120$$
$$6! = 720$$

$$7! = 5{,}040$$
$$8! = 40{,}320$$
$$9! = 362{,}880$$
$$10! = 3{,}628{,}800$$
$$11! = 39{,}916{,}800$$
$$12! = 479{,}001{,}600$$

Basic Series

The following are sometimes called *basic series* or *simple series* because they proceed according to elementary rules.

Arithmetic series

An *arithmetic series* is a series (A) such that:

$$f[a_{(n+1)}] = a_n + d$$

$$\therefore$$

$$A = a_1 + (a_1 + d) + (a_1 + 2 \times d) + (a_1 + 3 \times d) + \ldots$$

where d is a constant called the *difference*. For example, if $a_1 = 5$ and $d = 2$, then:

$$A = 5 + 7 + 9 + 11 + 13 \ldots$$

Geometric series

A *geometric series* is a series (G) such that:

$$f[a_{(n+1)}] = a_n/r$$

$$\therefore$$
$$G = a_1 + (a_1/r) + (a_1/r^2) + (a_1/r^3) + \ldots$$

where r is a constant called the *ratio*. For example, if $a_1 = 3$ and $r = 2$, then:

$$G = 3 + 3/2 + 3/4 + 3/8 + 3/16 + \ldots$$

Harmonic series

A *harmonic series* is a series ($H = a_1 + a_2 + a_3 + \ldots + a_n + \ldots$) such that the series consisting of the reciprocal of each term is an arithmetic series (A):

$$1/f\,[a_{(n+1)}] = 1/(a_n + d)$$
$$\therefore$$
$$A = H^{-1} = 1/a_1 + 1/a_2 + 1/a_3 + \ldots + 1/a_n + \ldots$$

where d is a constant. For example, if $a_1 = 1$ and $d = 3$, then:

$$H = 1,\ 1/4,\ 1/7,\ 1/10,\ 1/13,\ \ldots$$

Power series

A *power series* is a series (P) such that the following equation holds for coefficients a_i (where i is a non-negative integer subscript) and a variable (x):

$$P = a_0 + a_1 \times x + a_2 \times x^2 + a_3 \times x^3 + \ldots + a_n \times x^n + \ldots$$

where $a_1, a_2, a_3, \ldots + a_n, \ldots$ is a sequence. For example, if the sequence of coefficients is 2, 4, 6, 8, ... then:

$$P = 2 + 4 \times x + 6 \times x^2 + 8 \times x^3 + \ldots$$

More Sophisticated Series

The following series are encountered in electronics. They involve more sophisticated patterns than the basic series. In all examples, n is a positive integer denoting the ordinal number of the term in the series. Thus, for the first term, $n = 1$; for the second term, $n = 2$, and so on.

Arithmetic-geometric series

An *arithmetic-geometric series* is a series (C) such that, for constants a and b and variable x:

$$C = a + (a + b) \times x + (a + 2 \times b) \times x^2 + (a + 3 \times b) \times x^3 + \ldots + [a + (n-1) \times b] \times x^{(n-1)}$$

Taylor series

A *Taylor series*, also known as a *Taylor expansion*, is a series (T) such that the following equation holds for a function (f), its successive derivatives $f', f'', f^{(3)}, \ldots f^{(n)}, \ldots$, a constant ($a$), and a variable ($x$):

$$T = f(a) + [(x - a) \times (f'(a))] + [(x - a)^2 \times (f''(a))]/2 \\ + [(x - a)^3 \times (f^{(3)}(a))]/3! + \ldots + [(x - a)^n \times (f^{(n)}(a))]/n! + \ldots$$

Maclaurin series

A *Maclaurin series* M is a Taylor series with $a = 0$. The following holds for a function (f), its successive derivatives $f', f'', f^{(3)}, \ldots f^{(n)}, \ldots$, and a variable ($x$):

$$M = f(0) + [x \times (f'(0))] + [x^2 \times f''(0)]/2 \\ + [x^3 \times (f^{(3)}(0))]/3! + \ldots + [x^n \times (f^{(n)}(0))]/n! + \ldots$$

Fourier series

A *Fourier series* represents a periodic function (F) having a period ($2L$), such that the following equation holds for a variable (x), a sequence ($a_1, a_2, a_3, \ldots + a_n, \ldots$), and a sequence ($b_1, b_2, b_3, \ldots + b_n, \ldots$):

$$F = a_0/2 + a_1 \times \cos(\pi \times x/L) + b_1 \times \sin(\pi \times x/L) \\ + a_2 \times \cos(2 \times \pi \times x/L) + b_2 \times \sin(2 \times \pi \times x/L)$$

$$+ a_3 \times \cos(3 \times \pi \times x/L) + b_3 \times \sin(3 \times \pi \times x/L) + \ldots$$
$$+ a_n \times \cos(n \times \pi \times x/L) + b_n \times \sin(n \times \pi \times x/L) + \ldots$$

Trigonometric Series

Some circular and hyperbolic trigonometric functions can be represented as infinite series. Common examples are:

$$\sin x = x - x^3/3! + x^5/5! - x^7/7! + \ldots$$
$$+ (-1)^n \times [x^{(2n+1)} / (2n+1)!] + \ldots$$
$$\sinh x = x + x^3/3! + x^5/5! + x^7/7! + \ldots$$
$$+ x^{(2n+1)}/(2n+1)! + \ldots$$
$$\cos x = 1 - x^2/2! + x^4/4! - x^6/6! + \ldots$$
$$+ (-1)^n \times [x^{2n} / (2n)!]$$
$$\cosh x = 1 + x^2/2! + x^4/4! + x^6/6! + \ldots$$
$$+ x^{2n} / (2n)!$$

Logarithmic and Exponential Series

Some logarithmic and exponential functions can be represented as infinite series. Common examples are:

$$\ln(1+x) = x - x^2/2 + x^3/3 - x^4/4 + \ldots$$
$$+ (-1)^{(n-1)} \times (x^n/n) + \ldots$$

$$\ln(1-x) = x - x^2/2 - x^3/3 - x^4/4 - \ldots - x^n/n$$

$$\ln[(1+x)/(1-x)] = \\ 2 \times x + 2 \times x^3/3 + 2 \times x^5/5 + \ldots + \\ 2 \times x^{(2n-1)}/(2 \times n - 1)$$

$$e^x = 1 + x + x^2/2! + x^3/3! + \ldots \\ + x^n/n! + \ldots$$

$$e^{-x} = 1 - x + x^2/2! - x^3/3! + \ldots \\ + (-1)^n \times (x^n/n!) + \ldots$$

$$e^{jx} = 1 + jx - x^2/2! - jx^3/3! + \ldots \\ + (j^n) \times (x^n/n!) + \ldots$$

$$e^{-jx} = 1 - jx - x^2/2! + jx^3/3! + \ldots \\ + (-j)^n \times (x^n/n!) + \ldots$$

where j represents the j operator, equal to the positive square root of -1. (Note: In pure mathematics, this operator is denoted i. Also, when j is multiplied by a single-character variable or by a real number, it is customary to eliminate the multiplication symbol. So, for example, jx is written rather than $j \times x$, and $-j4.555$ rather than $-j \times 4.555$ or $j \times -4.555$).

Trigonometric/Exponential Formulas

Several formulas, derived from series expansions, relate trigonometric and exponential functions. Common examples are:

$$e^{jx} = \cos x + j \times \sin x$$
$$e^{-jx} = \cos x - j \times \sin x$$
$$e^{(a+jb)} = e^a \times (\cos b + j \times \sin b)$$
$$e^{(a-jb)} = e^a \times (\cos b - j \times \sin b)$$
$$\sin x = (e^{jx} - e^{-jx})/2$$
$$\sinh x = (e^x - e^{-x})/2$$
$$\cos x = (e^{jx} + e^{-jx})/2$$
$$\cosh x = (e^x + e^{-x})/2$$

Chapter 6

Sets, Functions, and Vectors

This chapter outlines definitions and formulas relevant to functions involving sets and vectors.

Sets

A *set* is a collection or group of definable *elements*. In electronics, set elements commonly include:

- Points on a line

- Instants in time
- Coordinates in a plane
- Coordinates on a display
- Coordinates in space
- Curves on a graph or display
- Digital logic states
- Locations in memory or storage
- Data bits, bytes, or characters
- Subscribers to a network

If an element (a) is contained in a set (A), then this fact is written as:

$$a \in A$$

Set intersection

The *intersection* of two sets A and B, written A $\cap$ B, is the set (C) such that the following statement is true for every element x:

$$x \in C \leftrightarrow x \in A \text{ and } x \in B$$

Set union

The *union* of two sets A and B, written A ∪ B, is the set (C) such that the following statement is true for every element (x):

$$x \in C \leftrightarrow x \in A \text{ or } x \in B$$

Subsets

A set (A) is a *subset* of a set (B), written A⊆B, if and only if the following holds true:

$$x \in A \rightarrow x \in B$$

Proper subsets

A set (A) is a *proper subset* of a set (B), written A ⊂ B, if and only if the following both hold true:

$$x \in A \rightarrow x \in B$$
$$A \neq B$$

Disjoint sets

Two sets (A and B) are *disjoint* if and only if all three of the following conditions are met:

$$A \neq \emptyset$$

Chapter Six

$$B \neq \emptyset$$
$$A \cap B = \emptyset$$

where $\emptyset$ denotes the *empty set*, also called the *null set*.

Coincident sets

Two nonempty sets (A and B) are coincident, if and only if, for all elements x:

$$x \in A \leftrightarrow x \in B$$

Venn diagrams

The *set functions* $\cap$ and $\cup$ can be represented by Venn diagrams. Figure 6.1 is a Venn diagram depicting set intersection. Figure 6.2 shows set union.

Functions

A *function* is a rule that defines a correspondence or relationship among the points in two different sets.

One-one function

Let A and B be two nonempty sets. Suppose that for every member of A, a function (f) assigns some

Sets, Functions, and Vectors 101

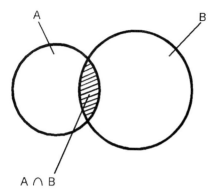

Figure 6.1 Intersection of nondisjoint, noncoincident sets A and B.

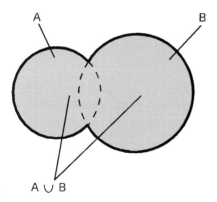

Figure 6.2 Union of nondisjoint, noncoincident sets A and B.

member of B. Let a_1 and a_2 be members of A. Let b_1 and b_2 be members of B, such that f assigns $f(a_1) = b_1$ and $f(a_2) = b_2$. Then f is a *one-one function* if and only if:

$$a_1 \neq a_2 \rightarrow b_1 \neq b_2$$

Onto function

A function (f) from set A to set B is an *onto function* if and only if:

$$b \in B \rightarrow f(a) = b \text{ for some } a \in A$$

One-to-one correspondence

A function (f) from set A to set B is a *one-to-one correspondence*, also known as a *bijection*, if and only if f is both one-one and onto.

Domain

Let f be a function from set A to set B. Let A' be the set of all elements (a) in A for which there is a corresponding element (b) in B. Then A' is called the *domain* of f. See Figures 6.3 and 6.4.

Range

Let f be a function from set A to set B. Let B' be the set of all elements (b) in B for which there is a

Sets, Functions, and Vectors 103

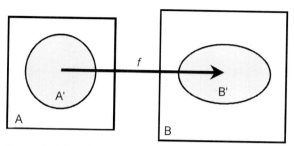

Figure 6.3 A function (f) from set A to set B, showing the domain (A') and the range (B').

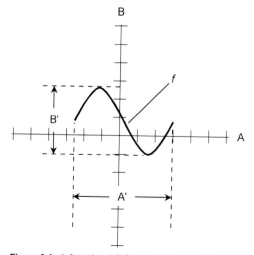

Figure 6.4 A function (f) from set A to set B in rectangular coordinates, showing the domain (A') and the range (B').

corresponding element (*a*) in A. Then B' is called the *range* of *f*. See Figures 6.3 and 6.4.

Continuity

A function (*f*) is *continuous* if and only if, for every point (*a*) in the domain A' and for every point (*b*) = *f*(*a*) in the range B', *f*(*x*) approaches *b* as *x* approaches *a*. If this requirement is not met for every point (*a*) in A', then the function (*f*) is *discontinuous*, and each point or value (a_d) in A' for which the requirement is not met is called a *discontinuity*. Examples of continuous and discontinuous functions are shown in Figure 6.5.

Linear function

A *linear function* of a variable *x* is a function (*f*) whose graph appears as a straight line in rectangular coordinates. Such a function always takes the form:

$$f(x) = a \times x + b$$

where *a* and *b* are constants.

Quadratic function

A *quadratic function*, also called a *second-order function*, of a variable (*x*) is a function (*f*) whose

Sets, Functions, and Vectors 105

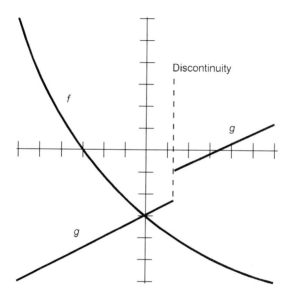

Figure 6.5 Function *f* is continuous; function *g* is discontinuous.

graph appears as a parabola in rectangular coordinates. Quadratic functions take the form:

$$f(x) = a \times x^2 + b \times x + c$$

where a, b, and c are constants.

Cubic function

A *cubic function*, also called a *third-order function*, of a variable (x) is a function (f) of the form:

$$f(x) = a \times x^3 + b \times x^2 + c \times x + d$$

where a, b, c, and d are constants.

Quartic function

A *quartic function*, also called a *fourth-order function*, of a variable (x) is a function (f) of the form:

$$f(x) = a \times x^4 + b \times x^3 + c \times x^2 + d \times x + e$$

where a, b, c, d, and e are constants.

Function of the *n*th order

An *nth-order function* of a variable (x) takes the general form:

$$f(x) = a_n \times x^n + a_{(n-1)} \times x^{(n-1)} + \ldots + a_2 \times x^2 + a_1 x + a_0$$

where a_0 through a_n are constants.

Generalized logarithmic function

A *logarithmic function* of a variable (x) takes the general form:

$$f(x) = a \times \log_n (b \times x) + c$$

where a, b, and c are constants, and n is the logarithm base. The two most common logarithmic bases are 10 and the exponential constant $e \approx 2.71828$.

Generalized exponential function

An *exponential function* of a variable x takes the general form:

$$f(x) = a \times n^{(b \times x)} + c$$

where a, b, and c are constants, and n is the exponential base. The two most common exponential bases are 10 and exponential constant $e \approx 2.71828$.

Generalized trigonometric function

The six fundamental *trigonometric functions* are outlined in Chapter 4. The general form for a trigonometric function (f) is:

$$f(x) = a \times \text{trig}\,(b \times x) + c$$

where *trig* represents the sine, cosine, tangent, cosecant, secant, or cotangent functions; and a, b, and c are constants.

Vectors

A *vector* is a mathematical expression for a quantity exhibiting two independently variable properties: *magnitude* and *direction*.

Vectors in the *XY*-plane

In the xy-plane, vectors **a** and **b** can be denoted as rays from the origin (0,0) to points (x_a, y_a) and (x_b, y_b), as shown in Figure. 6.6.

The magnitude of **a**, written |**a**|, is given by:

$$|\mathbf{a}| = (x_a^2 + y_a^2)^{1/2}$$

The direction of **a** is the angle θ_a that **a** subtends counterclockwise from the positive x axis:

$$\text{dir } \mathbf{a} = \theta_a = \arctan(y_a/x_a) = \tan^{-1}(y_a/x_a)$$

By convention, the following restrictions hold:

$$0 \leq \theta_a < 360 \text{ for } \theta_a \text{ in degrees}$$
$$0 \leq \theta_a < 2\pi \text{ for } \theta_a \text{ in radians}$$

The sum of vectors **a** and **b** is:

$$\mathbf{a} + \mathbf{b} = [(x_a + x_b), (y_a + y_b)]$$

This sum can be found geometrically by constructing a parallelogram with **a** and **b** as adjacent sides; then **a** + **b** is the diagonal of this parallelogram.

Sets, Functions, and Vectors 109

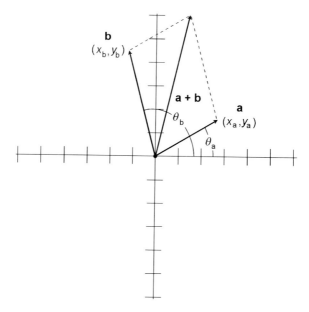

Figure 6.6 Vectors in the xy-plane.

The *dot product*, also known as the *scalar product* and written **a** • **b**, of vectors **a** and **b** is a real number given by the formula:

$$\mathbf{a} \bullet \mathbf{b} = x_a \times x_b + y_a \times y_b$$

The *cross product*, also known as the *vector product* and written $\mathbf{a} \times \mathbf{b}$, of vectors **a** and **b** is a vector perpendicular to the xy-plane and whose magnitude is given by the formula:

$$|\mathbf{a} \times \mathbf{b}| = |\mathbf{a}| \times |\mathbf{b}| \times \sin{[\tan^{-1}(y_b/x_b - y_a/x_a)]}$$

If the direction angle (θ_b) is greater than the direction angle (θ_a) (as shown in Figure 6.6), then $\mathbf{a} \times \mathbf{b}$ points toward the observer. If $\theta_b < \theta_a$, then $\mathbf{a} \times \mathbf{b}$ points away from the observer.

Vectors in the polar plane

In the polar coordinate plane, vectors **a** and **b** can be denoted as rays from the origin (0,0) to points (r_a, θ_a) and (r_b, θ_b), as shown in Figure 6.7.

The magnitude and direction of a vector (**a**) in the polar coordinate plane are defined directly:

$$|\mathbf{a}| = r_a$$
$$\text{dir } \mathbf{a} = \theta_a$$

By convention, the following restrictions hold:

$$r \geq 0$$
$$0 \leq \theta_a < 360 \text{ for } \theta_a \text{ in degrees}$$
$$0 \leq \theta_a < 2\pi \text{ for } \theta_a \text{ in radians}$$

The sum of **a** and **b** is best found by converting into rectangular (xy-plane) coordinates, adding the vectors according to the formula for the xy-plane,

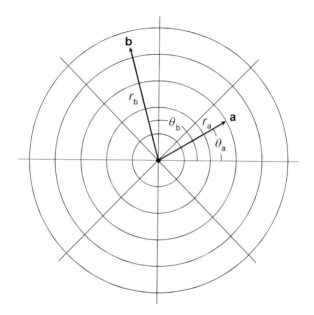

Figure 6.7 Vectors in the polar plane.

and then converting the resultant back to polar coordinates. To convert any vector (**a**) from polar to rectangular coordinates, these formulas apply:

$$x_a = r_a \times \cos \theta_a$$
$$y_a = r_a \times \sin \theta_a$$

To convert any vector **a** from rectangular coordinates to polar coordinates, these formulas apply:

$$r_a = (x_a^2 + y_a^2)^{1/2}$$
$$\theta_a = \arctan(y_a/x_a) = \tan^{-1}(y_a/x_a)$$

The dot product of **a** and **b** in polar coordinates is given by:

$$\mathbf{a} \bullet \mathbf{b} = |\mathbf{a}| \times |\mathbf{b}| \times \cos(\theta_b - \theta_a)$$

The cross product of **a** and **b** is perpendicular to the polar plane. Its magnitude is given by:

$$|\mathbf{a} \times \mathbf{b}| = |\mathbf{a}| \times |\mathbf{b}| \times \sin(\theta_b - \theta_a)$$

If $\theta_b > \theta_a$ (as is the case in Figure 6.7), then **a** points toward the observer. If $\theta_b < \theta_a$, then **a** points away from the observer.

Vectors in *xyz*-space

In rectangular xyz-space, vectors **a** and **b** can be denoted as rays from the origin (0,0,0) to points (x_a, y_a, z_a) and (x_b, y_b, z_b). The magnitude of **a**, written |**a**|, is given by:

$$|\mathbf{a}| = (x_a^2 + y_a^2 + z_a^2)^{1/2}$$

The direction of **a** is denoted by measuring the angles θ_x, θ_y, and θ_z that **a** subtends relative to the

Sets, Functions, and Vectors

x, y, and z axes, respectively. These angles, expressed in radians as an ordered triple $(\theta_x, \theta_y, \theta_z)$, are the *direction angles* of **a**. Often, the cosines of these angles are specified. These are the *direction cosines* of **a**:

$$\text{dir } \mathbf{a} = (\alpha, \beta, \gamma)$$
$$\alpha = \cos \theta_x$$
$$\beta = \cos \theta_y$$
$$\gamma = \cos \theta_z$$

The sum of vectors **a** and **b** is:

$$\mathbf{a} + \mathbf{b} = [(x_a + x_b), (y_a + y_b), (z_a + z_b)]$$

This sum can, as in the two-dimensional case, be found geometrically by constructing a parallelogram with **a** and **b** as adjacent sides. The sum **a + b** is the diagonal.

The dot product (**a • b**) of two vectors **a** and **b** in xyz-space is a real number given by the formula:

$$\mathbf{a} \bullet \mathbf{b} = x_a \times x_b + y_a \times y_b + z_a \times z_b$$

The cross product **a × b** of vectors **a** and **b** in xyz-space is a vector perpendicular to the plane P containing both **a** and **b**, and whose magnitude is given by the formula:

$$|\mathbf{a} \times \mathbf{b}| = |\mathbf{a}| \times |\mathbf{b}| \times \sin \theta_{ab},$$

where θ_{ab} is the angle between **a** and **b** as measured in P. Vector $\mathbf{a} \times \mathbf{b}$ is perpendicular to P. If **a** and **b** are observed from some point on a line perpendicular to P and intersecting P at the origin, and θ_{ab} is expressed counterclockwise from **a** to **b**, then **a** points toward the observer. If **a** and **b** are observed from some point on a line perpendicular to P and intersecting P at the origin, and θ_{ab} is expressed clockwise from **a** to **b**, then $\mathbf{a} \times \mathbf{b}$ points away from the observer.

Chapter 7

Differentiation and Integration

This chapter contains definitions and formulas relevant to basic differential and integral calculus.

Derivatives

Let f be a real-number function, x_0, an element of the domain of f, and y_0, an element of the range of f such that $y_0 = f(x_0)$. Suppose that f is continuous in the vicinity of (x_0, y_0), as shown in Figure 7.1.

Chapter Seven

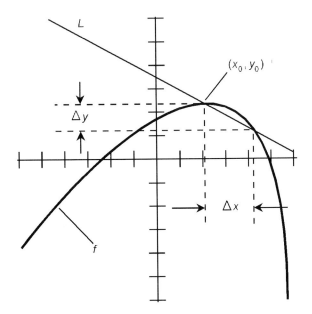

Figure 7.1 Derivative represented by the slope of a curve at a point.

Let Δx represent a small change in x, and let Δy represent the change in $y = f(x)$ that occurs as a result of Δx. Then the *derivative* of f at (x_0, y_0) is defined as:

$$f'(x_0) = \text{Lim}_{\Delta x \to 0}(\Delta y/\Delta x)$$

If f is continuous at all points (x_0) in its domain, then the derivative of f is defined and can be denoted in several ways:

$$f'(x) = d/dx \, (f) = df/dx = dy/dx$$

In Figure 7.1, the slope of line L approaches $f'(x_0)$ as $\Delta x \to 0$. For this reason, the derivative $f'(x_0)$ is graphically described as the slope of a line tangent to the curve of f at the point (x_0, y_0).

Second derivative

The *second derivative* of a function (f) is the derivative of its derivative. This can be denoted in various ways:

$$f''(x) = d^2/dx^2 \, (f) = d^2f/dx^2 = d^2y/dx^2$$

Higher-order derivatives

The *nth derivative* of a function (f) is the derivative taken in succession n times, where n is a positive integer. This can be denoted as follows:

$$f^{(n)}(x) = d^n/dx^n \, (f) = d^n f/dx^n = d^n y/dx^n$$

Sum and difference of derivatives

Let f and g be two different functions, and let $f + g = f(x) + g(x)$ for all x in the domains of both f and g. Then:

$$d(f + g)/dx = df/dx + dg/dx$$

and

$$d(f - g)/dx = df/dx - dg/dx$$

Multiplication by a constant

Let f be a function, let x be an element of the domain of f, and let c be a constant. Then:

$$d(c \times f)/dx = c \times (df/dx)$$

Product of derivatives

Let f and g be two different functions, and define $f \times g = f(x) \times g(x)$ for all x in the domains of both f and g. Then:

$$d(f \times g)/dx = f \times (dg/dx) + g \times (df/dx)$$

Quotient of derivatives

Let f and g be two different functions, and define $f/g = f(x)/g(x)$ for all x in the domains of both f and g. Then:

$$d(f/g)/dx = [g \times df/dx) - f \times (dg/dx)]/g^2$$

where $g^2 = [g(x) \times g(x)]$, not to be confused with $d^2 g/dg^2$.

Function raised to a power

Let f be a function, let x be an element of the domain of f, and let n be a positive integer. Then:

$$d(f^n)/dx = n(f^{n-1}) \times df/dx$$

where f^n denotes f multiplied by itself n times (not to be confused with the nth derivative).

Table of derivatives

Table 7.1 lists some commonly encountered functions and their derivatives.

TABLE 7.1 Derivatives. Letters *a*, *b*, and *c* Denote Constants. Letters *f*, *g*, and *h* Denote Functions; *m*, *n*, and *p* Denote Integers; *w*, *x*, *y*, and *z* Denote Variables. The Letter *e* Represents the Exponential Constant (Approximately 2.71828).

Function	Derivative
$f(x) = a$	$f'(x) = 0$
$f(x) = a \times x$	$f'(x) = a$
$f(x) = a \times x^n$	$f'(x) = n \times a \times x^{n-1}$
$f(x) = 1/x$	$f'(x) = \ln |x|$
$f(x) = \ln x$	$f'(x) = 1/x$

TABLE 7.1 *(Continued)*

Function	Derivative
$f(x) = \ln g(x)$	$f'(x) = g^{-1}(x) \times g'(x)$
$f(x) = 1/x^a$	$f'(x) = -a / (x^{a+1})$
$f(x) = e^x$	$f'(x) = e^x$
$f(x) = a^x$	$f'(x) = a^x \times \ln a$
$f(x) = a^{g(x)}$	$f'(x) = a^{g(x)} \times \ln a \times g'(x)$
$f(x) = e^{ax}$	$f'(x) = a \times e^x$
$f(x) = e^{g(x)}$	$f'(x) = e^{g(x)} \times g'(x)$
$f(x) = \sin x$	$f'(x) = \cos x$
$f(x) = \cos x$	$f'(x) = -\sin x$
$f(x) = \tan x$	$f'(x) = \sec^2 x$
$f(x) = \csc x$	$f'(x) = -\csc x \times \cot x$
$f(x) = \sec x$	$f'(x) = -\csc x - \tan x$
$f(x) = \cot x$	$f'(x) = -\csc^2 x$
$f(x) = \arcsin x = \sin^{-1} x$	$f'(x) = 1 / (1 - x^2)^{1/2}$
$f(x) = \arccos x = \cos^{-1} x$	$f'(x) = -1 / (1 - x^2)^{1/2}$
$f(x) = \arctan x = \tan^{-1} x$	$f'(x) = 1 / (1 + x^2)$
$f(x) = \text{arccsc } x = \csc^{-1} x$	$f'(x) = -1 / [x \times (x^2 - 1)^{1/2}]$
$f(x) = \text{arcsec } x = \sec^{-1} x$	$f'(x) = 1 / [x \times (x^2 - 1)^{1/2}]$
$f(x) = \text{arccot } x = \cot^{-1} x$	$f'(x) = -1 / (1 + x^2)$
$f(x) = \sinh x$	$f'(x) = \cosh x$
$f(x) = \cosh x$	$f'(x) = \sinh x$
$f(x) = \tanh x$	$f'(x) = \text{sech}^2 x$
$f(x) = \text{csch } x$	$f'(x) = -\text{csch } x \times \coth x$
$f(x) = \text{sech } x$	$f'(x) = \text{sech } x \times \tanh x$

TABLE 7.1 *(Continued)*

Function	Derivative
$f(x) = \coth x$	$f'(x) = -\operatorname{csch}^2 x$
$f(x) = \operatorname{arcsinh} x = \sinh^{-1} x$	$f'(x) = 1 / (x^2 + 1)^{1/2}$
$f(x) = \operatorname{arccosh} x = \cosh^{-1} x$	$f'(x) = 1 / (x^2 - 1)^{1/2}$
$f(x) = \operatorname{arctanh} x = \tanh^{-1} x$	$f'(x) = 1 / (1 - x^2)$
$f(x) = \operatorname{arccsch} x = \operatorname{csch}^{-1} x$	$f'(x) = -1 / [x \times (1 + x^2)^{1/2}]$ for $x > 0$
	$f'(x) = 1 / [x \times (1 + x^2)^{1/2}]$ for $x < 0$
$f(x) = \operatorname{arcsech} x = \operatorname{sech}^{-1} x$	$f'(x) = -1 / [x \times (1 - x^2)^{1/2}]$ for $x > 0$
	$f'(x) = 1 / [x \times (1 - x^2)^{1/2}]$ for $x < 0$
$f(x) = \operatorname{arccoth} x = \coth^{-1} x$	$f'(x) = 1 / (1 - x^2)$

Waveform derivatives

Figures 7.2 through 7.6 show several common waveforms and the results of their passing through a *differentiator circuit*.

Integrals

In general, *integration* is the opposite of differentiation. *Integral calculus* is used to find areas, volumes, and accumulated quantities.

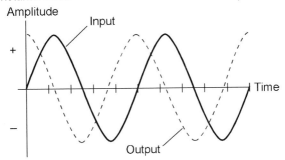

Figure 7.2 Results of passing a sine wave through a differentiator.

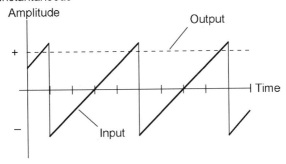

Figure 7.3 Results of passing an up-ramp wave through a differentiator. Negative transitions are assumed to occur instantaneously.

Differentiation and Integration 123

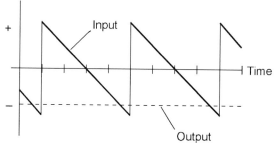

Figure 7.4 Results of passing a down-ramp wave through a differentiator. Positive transitions are assumed to occur instantaneously.

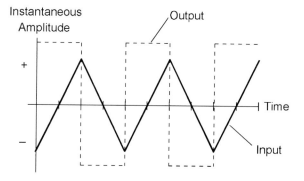

Figure 7.5 Results of passing a triangular wave through a differentiator.

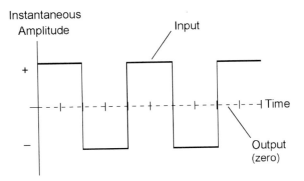

Figure 7.6 Results of passing a square wave through a differentiator. Positive and negative transitions are assumed to occur instantaneously.

Indefinite integral

Let $f(x)$ be a defined and continuous real-number function of a variable (x). The *antiderivative* or *indefinite integral* of f is a function (F) such that $dF/dx = f$. This is written as:

$$\int f(x)\, dx = F(x) + c$$

where c is a real number and dx is the differential of x (customarily annotated with all indefinite integrals).

Constant of integration

An infinite number of antiderivatives exist for any given function, all of which differ by real-number values. If the function $F_a(x)$ is an antiderivative of $f(x)$, then $F_b(x) = F_a(x) + c$ is also an antiderivative of $f(x)$, and c is known as the *constant of integration*.

Definite integral

Let $f(x)$ be a function that is defined and is continuous over the interval from $x = a$ to $x = b$. Let F be any antiderivative of f. The *definite integral* from a to b is:

$$F(a) - F(b)$$

The constant of integration becomes irrelevant in the definite integral because it is subtracted from itself. The definite integral can be depicted as the area under the curve of f in rectangular coordinates (Figure 7.7). Areas above the x axis are considered positive; "areas" below the x axis are considered negative.

Linearity

Let f and g be defined, continuous functions of x. Let a and b be real-number constants. Then:

$$\int [a \times f(x) + b \times g(x)]dx = a \times \int f(x)\, dx$$
$$+ b \times \int g(x)\, dx$$

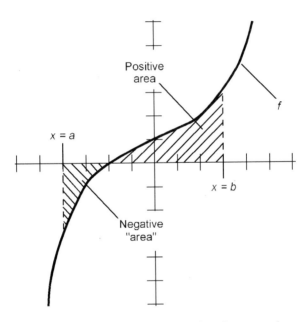

Figure 7.7 Definite integral represented as the area under a curve between two points.

Integration by parts

Let f and g be defined, continuous functions of x, and let F be an antiderivative of f. Then:

$$\int [f(x) \times g(x)]\, dx = F(x) \times g(x) - \int [F(x) \times dg/dx]\, dx$$

Table of indefinite integrals

Table 7.2 lists some commonly encountered functions and their indefinite integrals.

TABLE 7.2 Indefinite Integrals. Letters *a*, *b*, and *c* Denote Constants; *f*, *g*, and *h* Denote Functions; *m*, *n*, and *p* Denote Integers; *w*, *x*, *y*, and *z* Denote Variables. The Letter *e* Represents the Exponential Constant (Approximately 2.71828).

Function	Derivative
$f(x) = 0$	$\int f(x)\, dx = c$
$f(x) = 1$	$\int f(x)\, dx = 1 + c$
$f(x) = a$	$\int f(x)\, dx = a + c$
$f(x) = x$	$\int f(x)\, dx = x^2/2 + c$
$f(x) = a \times x$	$\int f(x)\, dx = a \times x^2/2 + c$

TABLE 7.2 (Continued)

Function	Derivative
$f(x) = a \times x^2$	$\int f(x)\,dx = a \times x^3/3 + c$
$f(x) = a \times x^3$	$\int f(x)\,dx = a \times x^4/4 + c$
$f(x) = a \times x^4$	$\int f(x)\,dx = a \times x^5/5 + c$
$f(x) = a/x = a \times x^{-1}$	$\int f(x)\,dx = a \times \ln|x| + c$
$f(x) = a/x^2 = a \times x^{-2}$	$\int f(x)\,dx = -a/x + c$
$f(x) = a/x^3 = a \times x^{-3}$	$\int f(x)\,dx = -a/(2 \times x^2) + c$
$f(x) = a/x^4 = a \times x^{-4}$	$\int f(x)\,dx = -a/(3 \times x^3) + c$
$f(x) = a \times x^n$	$\int f(x)\,dx = [a \times x^{n+1}/(n+1)] + c$ for $n \neq -1$
$f(x) = a \times g(x)$	$\int f(x)\,dx = a \times \int g(x)\,dx + c$
$f(x) = g(x) + h(x)$	$\int f(x)\,dx = \int g(x)\,dx + \int h(x)\,dx + c$
$f(x) = h(x) \times g'(x)$	$\int f(x)\,dx = g(x) \times h(x) - \int g(x) \times h'(x) + c$
$f(x) = e^x$	$\int f(x)\,dx = e^x + c$
$f(x) = a \times e^{bx}$	$\int f(x)\,dx = a \times e^x/b + c$

Differentiation and Integration

TABLE 7.2 (Continued)

Function	Derivative
$f(x) = \ln x$	$\int f(x)dx = x \times \ln x - x + c$
$f(x) = \sin x$	$\int f(x)dx = -\cos x + c$
$f(x) = \cos x$	$\int f(x)dx = \sin x + c$
$f(x) = \tan x$	$\int f(x)dx = \ln |\sec x| + c$
$f(x) = \csc x$	$\int f(x)dx = \ln |\tan (x/2)| + c$
$f(x) = \sec x$	$\int f(x)dx = \ln |\sec x + \tan x| + c$
$f(x) = \cot x$	$\int f(x)dx = \ln |\sin x| + c$
$f(x) = \sec x \times \tan x$	$\int f(x)\,dx = \sec x + c$
$f(x) = \sin^2 x$	$\int f(x)\,dx = \{x - [\sin (2 \times x)]/2\}/2 + c$
$f(x) = \cos^2 x$	$\int f(x)\,dx = \{x + [\sin (2 \times x)]/2\}/2 + c$
$f(x) = \sinh x$	$\int f(x)\,dx = \cosh x + c$
$f(x) = \cosh x$	$\int f(x)\,dx = \sinh x + c$
$f(x) = \tanh x$	$\int f(x)\,dx = \ln |\cosh x| + c$
$f(x) = \csch x$	$\int f(x)\,dx = \ln |\tanh (x/2)| + c$

TABLE 7.2 *(Continued)*

Function	Derivative
$f(x) = \text{sech } x$	$\int f(x)\, dx = 2 \times \arctan(e^x) + c$ $= 2 \times \tan^{-1}(e^x) + c$
$f(x) = \coth x$	$\int f(x)\, dx = \ln|\sinh x| + c$

Waveform integrals

Figures 7.8 through 7.12 show several common waveforms and the results of their passing

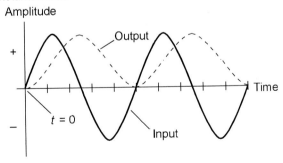

Figure 7.8 Results of passing a sine wave through an integrator.

Differentiation and Integration 131

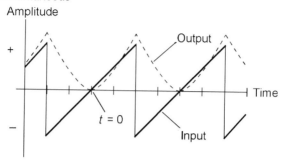

Figure 7.9 Results of passing an up-ramp wave through an integrator. Negative transitions are assumed to occur instantaneously.

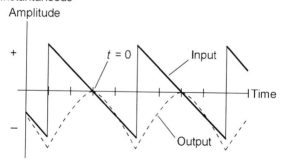

Figure 7.10 Results of passing a down-ramp wave through an integrator. Positive transitions are assumed to occur instantaneously.

132 Chapter Seven

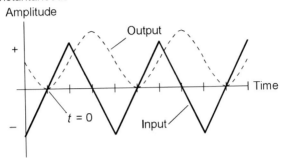

Figure 7.11 Results of passing a triangular wave through an integrator.

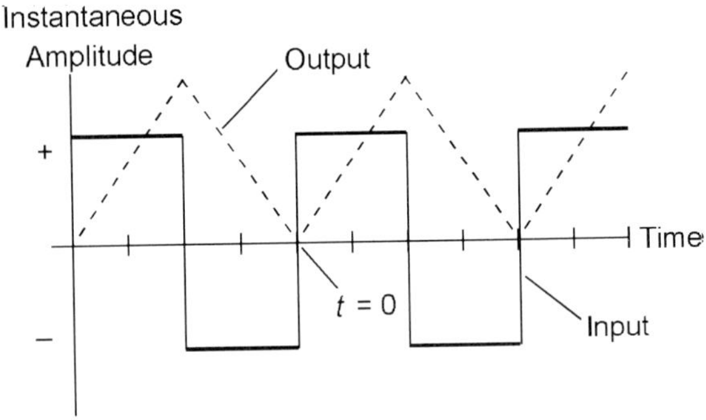

Figure 7.12 Results of passing a square wave through an integrator. Positive and negative transitions are assumed to occur instantaneously.

through an *integrator circuit*. The start time ($t = 0$) of the input wave cycle affects the output in some cases. The constant of integration, which appears as a direct current superimposed on the output wave, depends on the point assigned $t = 0$. These illustrations give specific examples for $t = 0$, as shown.

Chapter 8

Direct Current

This chapter contains formulas involving direct-current (DC) charge quantity, amperage, voltage, resistance, power, and energy.

DC Charge

The standard unit of electrical charge quantity, symbolized by Q, is the *coulomb,* equivalent to the charge contained in approximately 6.24×10^{18} electrons.

Charge vs. current and time

Let Q represent charge quantity in coulombs, let I represent direct current in amperes, and let t represent time in seconds. Then:

$$Q = \int I \, dt$$

This principle is illustrated in Figure 8.1. If the current remains constant over time, then this formula can be simplified to:

$$Q = I \times t$$

Coulomb's law

Let F represent force in newtons, let Q_X and Q_Y represent the charges on two distinct objects X and Y, and let d represent the distance between the charge centers of X and Y. Then:

$$F = (Q_X \times Q_Y)/d^2$$

If the charges are alike in polarity (+/+ or –/–), then F is positive (repulsive). If the charges are opposite in polarity (–/+ or +/–), then F is negative (attractive).

DC Amperage

The standard unit of direct current, also called *DC amperage* and symbolized by I, is the *ampere*, equivalent to one coulomb of charge moving past

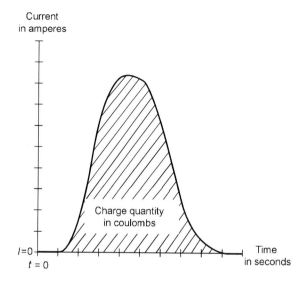

Figure 8.1 Electrical charge as a function of current and time.

a point in one second, always in the same direction.

Charging and discharging

Let $I_{c/d}$ represent instantaneous charging or discharging current in amperes. Let t represent time in hours. Then the accumulated charge or discharge quantity (Q_{Ah}), in ampere-hours, is:

$$Q_{Ah} = \int I_{c/d}\, dt$$

138   Chapter Eight

If the rate of charging or discharging is constant, then:

$$Q_{Ah} = I_{c/d} \times t$$

Current vs. charge and time

Let Q represent charge quantity in coulombs. Let t represent time in seconds. Then the instantaneous charging or discharging current ($I_{c/d}$), in amperes, is:

$$I_{c/d} = dQ/dt$$

If the rate of charge or discharge is constant over an interval beginning at time t_1 and ending at time t_2, then:

$$I_{c/d} = (Q_2 - Q_1)/(t_2 - t_1)$$

where Q_1 is the charge at time t_1, and Q_2 is the charge at time t_2. In these formulas, positive values of $I_{c/d}$ represent a charging condition; negative values of $I_{c/d}$ represent a discharging condition.

Ohm's law for DC amperage

Let V represent the voltage (in volts) across a component or device; let R represent the resistance (in ohms) of the component or device. Then the current I (in amperes) through the component or device is:

$$I = V/R$$

Current vs. voltage and power

Let V represent the potential difference (in volts) across a component or device; let P represent the power (in watts) dissipated, radiated, or supplied by the component or device. Then the current I (in amperes) through the component or device is:

$$I = P/V$$

Current vs. voltage, energy, and time

Let V represent the potential difference (in volts) across a component or device; let E represent the energy (in joules) dissipated, radiated, or supplied by the component or device over period of time t (in seconds). Then current I (in amperes) through the component or device is:

$$I = E/(V \times t)$$

Current vs. resistance and power

Let R represent the resistance (in ohms) of a component or device; let P represent the power (in watts) radiated, dissipated, or supplied by the component or device. Then the current I (in amperes) through the component or device is:

$$I = (P/R)^{1/2}$$

Current vs. resistance, energy, and time

Let R represent the resistance (in ohms) of a component or device; let E represent the energy (in joules) dissipated, radiated, or supplied by the component or device over period of time t (in seconds). Then current I (in amperes) through the component or device is:

$$I = [E/(R \times t)]^{1/2}$$

Kirchhoff's Law for DC Amperage

The current going into any point in a DC circuit is the same as the current going out. An example is shown at Figure 8.2. If I_{in} represents the total current entering the branch point (Z) and I_{out} represents the total current emerging from point Z, then:

$$I_{in} = I_{out}$$
$$I_{in} = I_1 + I_2$$
$$I_{out} = I_3 + I_4 + I_5$$
$$\therefore > I_1 + I_2 = I_3 + I_4 + I_5$$

DC Voltage

The standard unit of DC voltage, also called *potential difference* or *electromotive force* (EMF), is the *volt*. Voltage is symbolized by V in this chap-

Direct Current 141

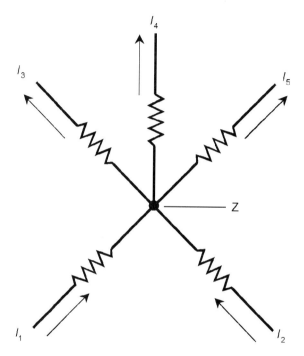

Figure 8.2 Kirchhoff's current law.

ter; alternatively it can be symbolized by E, unless doing so would confuse it with energy. One volt is the EMF required to drive one ampere of current through a resistance of one ohm.

Ohm's law for DC voltage

Let I be the current (in amperes) through a component, and let R be the resistance (in ohms) of that component. Then potential difference V (in volts) across the component is:

$$V = I \times R$$

Voltage vs. current and power

Let I represent the current (in amperes) through a component or device. Let P represent the power (in watts) dissipated, radiated, or supplied by the component or device. Then potential difference V (in volts) across the component or device is:

$$V = P/I$$

Voltage vs. current, energy, and time

Let I represent the current (in amperes) through a component or device. Let E represent the energy (in joules) dissipated, radiated, or supplied by the component or device over period of time t (in seconds). Then potential difference V (in volts) across the component or device is:

$$V = E/(I \times t)$$

Voltage vs. resistance and power

Let R represent the resistance (in ohms) of a component or device. Let P represent the power (in watts) dissipated, radiated, or supplied by the component or device. Then potential difference V (in volts) across the component or device is:

$$V = (P \times R)^{1/2}$$

Voltage vs. resistance, energy, and time

Let R represent the resistance (in ohms) of a component or device. Let E represent the energy (in joules) dissipated, radiated, or supplied by the component or device over period of time t (in seconds). Then potential difference V (in volts) across the component or device is:

$$V = (E \times R/t)^{1/2}$$

Kirchhoff's law for DC voltage

In a closed DC network, the sum of the voltages across all the components in any given loop, taking polarity into account, is zero. An example is shown in Figure 8.3. The EMF of the battery is V; there are four components across which the potential differences are V_1, V_2, V_3, and V_4. The following equations hold in this case:

144 Chapter Eight

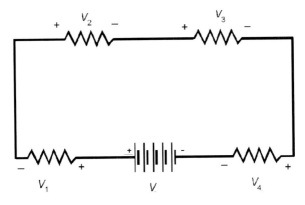

Figure 8.3 Kirchhoff's voltage law.

$$V + V_1 + V_2 + V_3 + V_4 = 0$$
$$V_1 + V_2 + V_3 + V_4 = -V$$
$$V = -(V_1 + V_2 + V_3 + V_4)$$

DC Resistance

The standard unit of DC resistance, symbolized by R, is the *ohm*. A component has a resistance of one ohm when an applied EMF of one volt across it results in a current of one ampere through it, or when a current of one ampere through the component produces a potential difference of one volt across it.

Ohm's law for DC resistance

Let I be the current (in amperes) through a component, and let V be the potential difference (in volts) across it. Then resistance R of the component (in ohms) is:

$$R = V/I$$

Resistance vs. current and power

Let I be the current (in amperes) through a component or device, and let P be the power (in watts) radiated or supplied. Then resistance R (in ohms) of the component or device is:

$$R = P/I^2$$

Resistance vs. current, energy, and time

Let I be the current (in amperes) through a component or device, and let E be the energy (in joules) dissipated, radiated, or supplied over period of time t (in seconds). Then resistance R (in ohms) of the component or device is:

$$R = E / (I^2 \times t)$$

Resistance vs. voltage and power

Let V be the potential difference (in volts) across a component or device, and let P be the power (in watts) dissipated, radiated, or supplied. Then resistance R (in ohms) of the component or device is:

$$R = V^2/P$$

Resistance vs. voltage, energy, and time

Let V be the potential difference (in volts) across a component or device, and let E be the energy (in joules) dissipated, radiated, or supplied over period of time t (in seconds). Then resistance R (in ohms) of the component or device is:

$$R = V^2 \times t/E$$

DC Power

The standard unit of DC power, symbolized by P, is the *watt*. A component dissipates, radiates, or supplies one watt when it carries or provides a current of one ampere, and when the potential difference across it is one volt.

Power vs. energy and time

Let E be the energy (in joules) dissipated, radiated, or supplied by a component or device over period of time t (in seconds). Then dissipated, radiated, or supplied power P (in watts) is:

$$P = E/t$$

Power vs. current and voltage

Let I be the current (in amperes) through a component or device, and let V be the potential difference (in volts) across it. Then dissipated, radiated, or supplied power P (in watts) is:

$$P = V \times I$$

Power vs. current and resistance

Let I be the current (in amperes) through a component or device, and let R be its resistance (in ohms). Then dissipated, radiated, or supplied power P (in watts) is:

$$P = I^2 \times R$$

Power vs. voltage and resistance

Let V be the potential difference (in volts) across a component or device, and let R be its resistance (in ohms). Then dissipated, radiated, or supplied power P (in watts) is:

$$P = V^2/R$$

DC Energy

The standard unit of DC energy, symbolized by E, is the *joule*. A component dissipates, radiates, or supplies one joule when it dissipates, radiates, or supplies an average power of one watt over a time interval of one second.

Energy vs. power and time

Let P be the power (in watts) dissipated, radiated, or supplied by a component or device over period of time t (in seconds). Then energy E (in joules) is:

$$E = \int P\, dt$$

If the power remains constant over the entire time interval, then:

$$E = P \times t$$

Energy vs. current, voltage, and time

Let I be the current (in amperes) through a component or device, and let V be the potential difference (in volts) across it. Then dissipated, radiated, or supplied energy E (in joules) over period of time t (in seconds) is:

$$E = \int (V \times I)\, dt$$

If the current and voltage remain constant over the entire time interval, then:

$$E = V \times I \times t$$

Energy vs. current, resistance, and time

Let I be the current (in amperes) through a component or device, and let R be its resistance (in ohms). Then dissipated or radiated energy E (in joules) over period of time t (in seconds) is:

$$E = \int (I^2 \times R)\, dt$$

If the current and resistance remain constant over the entire time interval, then:

$$E = I^2 \times R \times t$$

Energy vs. voltage, resistance, and time

Let V be the potential difference (in volts) across a component or device, and let R be its resistance (in ohms). Then dissipated or radiated energy E (in joules) over period of time t (in seconds) is:

$$E = \int (V^2/R)\, dt$$

If the current and resistance remain constant over the entire time interval, then:

$$E = V^2 \times t/R$$

Chapter 9

Alternating Current

This chapter contains formulas involving alternating-current (AC) frequency, phase, amperage, voltage, impedance, power, and energy.

Frequency and Phase

Frequency is usually symbolized by the letter f, period by the letter T, and phase angle by the Greek letter ϕ.

Frequency vs. period

Let f be the frequency of an AC wave (in hertz), and let T be the period (in seconds). Then the following relations hold:

$$f = 1/T$$
$$T = 1/f$$

These relations are also valid for T in milliseconds (ms) and f in kilohertz (kHz); for T in microseconds (µs) and f in megahertz (MHz); for T in nanoseconds (ns) and f in gigahertz (GHz); and for T in picoseconds (ps) and f in terahertz (THz).

Phase angle vs. time and frequency

Let f be the frequency of an AC wave (in hertz), and let t be the time (in seconds) following the instant (t_0) at which the wave amplitude is zero and positive-going (Figure 9.1). Then the phase angle (ϕ), in degrees, is:

$$\phi = 360 \times f \times t$$

If ϕ is expressed in radians, then:

$$\phi = 2 \times \pi \times f \times t$$

Alternating Current 153

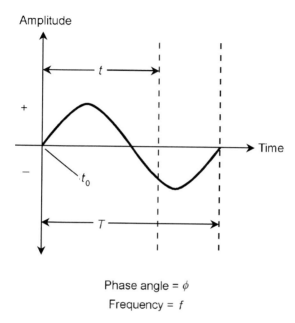

Phase angle = ϕ
Frequency = f

Figure 9.1 Relations among frequency (f), period (T), phase (ϕ), and time (t) for an AC sine wave cycle beginning at $t = t_0$.

These formulas are also valid for t in milliseconds (ms) and f in kilohertz (kHz); for t in microseconds (µs) and f in megahertz (MHz); for t in nanoseconds (ns) and f in gigahertz (GHz); and for t in picoseconds (ps) and f in terahertz (THz).

Phase angle vs. time and period

Let T be the period of an AC wave (in seconds) and let t be the time (in seconds) following the instant (t_0) at which the wave amplitude is zero and positive-going. Then the phase angle (ϕ), in degrees, is:

$$\phi = 360 \times t/T$$

If ϕ is expressed in radians, then:

$$\phi = 2 \times \pi \times t/T$$

These formulas are also valid for t and T in milliseconds (ms), microseconds (μs), nanoseconds (ns), and picoseconds (ps).

AC Amplitude Expressions

The amplitude of an AC wave can be expressed in several ways. The following formulas apply to sinusoidal waveforms and are expressed in terms of voltage (V). However, these formulas also apply to current (I).

Instantaneous amplitude

The *instantaneous amplitude* (V_{inst}) of an AC sine wave constantly varies. In the example at Figure 9.2, instantaneous amplitudes are shown as vertical displacement on the waveform.

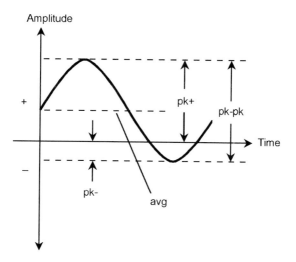

Figure 9.2 Positive peak (pk+), negative peak (pk−), peak-to-peak (pk-pk), and average (avg) values for an AC sine wave. See text for definition of root-mean-square (rms).

Positive peak amplitude

The *positive peak amplitude* (V_{pk+}) of an AC sine wave is the maximum deviation of V_{inst} in the positive direction. See Figure 9.2.

Negative peak amplitude

The *negative peak amplitude* (V_{pk-}) of an AC sine wave is the maximum deviation of V_{inst} in the negative direction. See Figure 9.2.

DC component

The *DC component* (V_{DC}) of an AC sine wave is the arithmetic mean of the positive and negative peak amplitudes:

$$V_{DC} = (V_{pk+} + V_{pk-})/2$$

Average amplitude

The *average amplitude* (V_{avg}) of an AC sine wave is the same as the DC component.

Peak amplitude when V_{DC} = 0

If $V_{DC} = 0$, then the positive and negative peak amplitudes are equal and opposite. This can be generalized as the *peak amplitude* (V_{pk}):

$$V_{pk} = V_{pk+} = -V_{pk-}$$

Alternating Current 157

Peak-to-peak amplitude

The *peak-to-peak amplitude* ($V_{\text{pk-pk}}$) of an AC wave is the difference between the positive and negative peak amplitudes:

$$V_{\text{pk-pk}} = V_{\text{pk+}} - V_{\text{pk-}}$$

If $V_{\text{DC}} = 0$, then:

$$V_{\text{pk-pk}} = 2 \times V_{\text{pk+}} = -2 \times V_{\text{pk-}}$$

Instantaneous amplitude vs. phase angle

Let $V_{\text{pk+}}$ represent the positive peak amplitude of a wave. Let V_{DC} represent the DC component, and let ϕ represent the phase angle, as measured from the point in time at which the instantaneous amplitude $V_{\text{inst}} = V_{\text{DC}}$ and is increasing positively. Then:

$$V_{\text{inst}} = V_{\text{DC}} + (V_{\text{pk+}} \times \sin \phi)$$

Effective amplitude

The *effective amplitude* of an AC wave is also known as the *direct-current* (DC) *equivalent amplitude* or the *root-mean-square* (*rms*) *amplitude*. Let V_{DC} represent the DC component. Then the rms amplitude, V_{rms}, is given by:

$$V_{\text{rms}} = V_{\text{DC}} + [2^{-1/2} \times (V_{\text{pk+}} - V_{\text{DC}})]$$
$$\approx V_{\text{DC}} + [0.707 \times (V_{\text{pk+}} - V_{\text{DC}})]$$

If there is no DC component, then:

$$V_{\mathrm{rms}} = 2^{-1/2} \times V_{\mathrm{pk}} \approx 0.707 \times V_{\mathrm{pk}}$$

Complex Numbers

In expressions of *complex impedance*, the positive square root of –1, known as the *j operator* and symbolized j, is used. The set of *imaginary numbers* derives from real-number multiples of j. In engineering, imaginary numbers and variables are written as j followed by a real number or variable. (Note: In pure mathematics, this operator is denoted i. Also, when j is multiplied by a single-character variable or by a real number, it is customary to eliminate the multiplication symbol. So, for example, jx is written, rather than $j \times x$, and $-j4.555$, rather than $-j \times 4.555$ or $j \times -4.555$.)

A complex impedance is the sum of a non-negative real number or real variable, plus a real-number or real-variable multiple of the j operator. Examples are:

$$8 + j5$$
$$3 + j0$$
$$5.355 - j0.677$$
$$a + jb$$
$$R + jX$$

Addition

Adding complex numbers requires adding the real-number and the imaginary-number parts separately:

$$(a + jb) + (c + jd) = (a + c) + j(b + d)$$

Subtraction

Subtracting complex numbers requires subtracting the real-number and the imaginary-number parts separately:

$$(a + jb) - (c + jd) = (a - c) + j(b - d)$$

Multiplication

The product of two complex numbers is given by:

$$(a + jb) \times (c + jd) = (a \times c - b \times d) + j(a \times d + b \times c)$$

Absolute value

The *absolute value* or *magnitude* of a complex number $(a + jb)$ is equal to the positive square root of the sum of the squares of its real-number coefficients:

$$|a + jb| = + (a^2 + b^2)^{1/2}$$

Impedance

Impedance, symbolized by the letter Z, is the opposition that a component or circuit offers to AC. Impedance is a two-dimensional quantity, consisting of two independent components, *resistance* and *reactance*.

Inductive reactance

Inductive reactance is symbolized jX_L. Its real-number coefficient, X_L, is always positive or zero.

jX_L vs. frequency

If the frequency of an AC source is given (in hertz) as f, and the inductance of a component is given (in henrys) as L, then the vector expression for inductive reactance (in imaginary-number ohms), jX_L, is:

$$jX_L = j\,(2 \times \pi \times f \times L) \approx j\,(6.28 \times f \times L)$$

This formula also applies for f in kilohertz (kHz) and L in millihenrys (mH); for f in megahertz (MHz) and L in microhenrys (µH); for f in gigahertz (GHz) and L in nanohenrys (nH); and for f in terahertz (THz) and L in picohenrys (pH).

RL phase angle

The phase angle ϕ_{RL} in a resistance-inductance (*RL*) circuit is the arctangent of the ratio of the real-number coefficient of the inductive reactance to the resistance:

$$\phi_{RL} = \tan^{-1}(X_L/R)$$

Capacitive reactance

Capacitive reactance is symbolized jX_C. Its real-number coefficient, X_C, is always negative or zero.

jX$_C$ vs. frequency

If the frequency of an AC source is given (in hertz) as f, and the value of a capacitor is given (in farads) as C, then the vector expression for capacitive reactance (in imaginary-number ohms), jX_C, is given by:

$$jX_C = -j\,[1/(2 \times \pi \times f \times C)] \approx -j\,[1/(6.28 \times f \times C)]$$

This formula also applies for f in megahertz (MHz) and C in microfarads (µF), and for f in terahertz (THz) and C in picofarads (pF).

RC phase angle

The phase angle ϕ_{RC} in a resistance-capacitance (*RC*) circuit is the arctangent of the ratio of the real-number coefficient of the capacitive reactance to the resistance:

$$\phi_{RC} = \tan^{-1}(X_C/R)$$

Complex impedances in series

Given two complex impedances $Z_1 = R_1 + jX_1$ and $Z_2 = R_2 + jX_2$ connected in series, the resultant complex impedance (Z) is their vector sum, given by:

$$Z = (R_1 + R_2) + j\,(X_1 + X_2)$$

Admittance

Admittance, symbolized by the letter Y, is the readiness with which a component or circuit conducts AC. Admittance is a two-dimensional quantity, consisting of two independent components, *conductance* and *susceptance*.

AC conductance

In an AC circuit, electrical conductance behaves the same way as in a DC circuit. Conductance is symbolized by the capital letter G. The relationship between conductance and resistance is:

$$G = 1/R$$

The unit of conductance is the *siemens*, sometimes called the *mho*.

Inductive susceptance

Inductive susceptance is symbolized jB_L. Its real-number coefficient, B_L, is always negative or zero, being the negative reciprocal of the real-number coefficient of inductive reactance:

$$B_L = -1/X_L$$

The vector expression of inductive susceptance requires the j operator, as does the vector expression of inductive reactance. The reciprocal of j is its negative, so when calculating vector quantities B_L in terms of vector quantities (X_L), the sign changes.

jB_L vs. frequency

If the frequency of an AC source is given (in hertz) as f, and the value of an inductor is given (in henrys) as L, then the vector expression for inductive susceptance (in imaginary-number siemens), jB_L, is given by:

$$jB_L = -j\,[1/(2 \times \pi \times f \times L)] \approx -j\,[1/(6.28 \times f \times L)]$$

This formula also applies for f in kilohertz (kHz) and L in millihenrys (mH); for f in megahertz (MHz) and L in microhenrys (µH); for f in gigahertz (GHz) and L in nanohenrys (nH); and for f in terahertz (THz) and L in picohenrys (pH).

Capacitive susceptance

Capacitive susceptance is symbolized jB_C. Its real-number coefficient, B_C, is always positive or zero, being the negative reciprocal of the real-number coefficient of capacitive reactance:

$$B_C = -1/X_C$$

The expression of capacitive susceptance requires the j operator, as does the expression of capacitive reactance. The reciprocal of j is its negative, so when calculating B_C in terms of X_C, the sign changes.

jB_C vs. frequency

If the frequency of an AC source is given (in hertz) as f, and the value of a capacitor is given (in farads) as C, then the vector expression for capacitive susceptance (in imaginary-number siemens), jB_C, is given by:

$$jB_C = j\,(2 \times \pi \times f \times C) \approx j\,(6.28 \times f \times C)$$

This formula also applies for f in megahertz (MHz) and C in microfarads (μF), and for f in terahertz (THz) and C in picofarads (pF).

Complex admittances in parallel

Given two complex admittances ($Y_1 = G_1 + jB_1$ and $Y_2 = G_2 + jB_2$) connected in parallel, the resultant complex admittance (Y) is their vector sum, given by:

$$Y = (G_1 + G_2) + j(B_1 + B_2)$$

Complex impedances in parallel

To find the resultant complex impedance of two components in parallel, follow these steps in order:

- Convert each real-number resistance to conductance: $G_n = 1/R_n$
- Convert each imaginary-number reactance to susceptance, paying careful attention to the changes in sign of the real-number coefficients: $B_n = -1/X_n$
- Sum the conductances and susceptances to get complex admittances

- Use the previous formula to find net complex admittance, consisting of a resultant conductance and a resultant susceptance
- Convert the resultant real-number conductance back to resistance
- Convert the resultant imaginary-number susceptance back to reactance, paying careful attention to the change in sign of the real-number coefficient: $X_n = -1/B_n$

The resulting expression $(R + jX)$ is the complex impedance of the two components in parallel.

AC Amperage

The standard unit of alternating current, also called *AC amperage* and symbolized by I_{rms}, is the *ampere rms*.

Current vs. voltage and reactance

Let V_{rms} be the AC voltage (in volts rms) across a component or device; let X be the real-number coefficient of the reactance (in ohms) of the component or device. Then the alternating current (in amperes rms), I_{rms}, is given by:

$$I_{rms} = |V_{rms}/X|$$

Current vs. voltage, frequency, and inductance

Let V_{rms} be the AC voltage (in volts rms) across a component or device; let f be the AC frequency (in hertz); let L be the inductance (in henrys) of the component or device. Then the alternating current (in amperes rms), I_{rms}, is given by:

$$I_{rms} = V_{rms}/(2 \times \pi \times f \times L) \approx V_{rms}/(6.28 \times f \times L)$$

This formula also applies for f in kilohertz (kHz) and L in millihenrys (mH); for f in megahertz (MHz) and L in microhenrys (µH); for f in gigahertz (GHz) and L in nanohenrys (nH); and for f in terahertz (THz) and L in picohenrys (pH).

Current vs. voltage and complex impedance

Let V_{rms} be the AC voltage (in volts rms) across a component or device; let f be the AC frequency (in hertz); let C be the capacitance (in farads) of the component or device. Then the alternating current (in amperes rms), I_{rms}, is given by:

$$I_{rms} = 2 \times \pi \times V_{rms} \times f \times C \approx 6.28 \times V_{rms} \times f \times C$$

This formula also applies for f in megahertz (MHz) and C in microfarads (µF), and for f in terahertz (THz) and C in picofarads (pF).

Current vs. voltage and complex impedance

Let V_{rms} be the AC voltage (in volts rms) across a component or device; let the complex impedance of the component or device be $Z = R + jX$, where X is the real-number coefficient of the reactance (in ohms) of the component or device and R is the resistance (in ohms) of the component or device. Then the alternating current (in amperes rms), I_{rms}, is given by:

$$I_{rms} = V_{rms}/(R^2 + X^2)^{1/2}$$

AC Voltage

The standard unit of AC voltage, also called *AC electromotive force* (AC EMF) and symbolized by V_{rms}, is the *volt rms*.

Voltage vs. current and reactance

Let I_{rms} be the alternating current (in amperes rms) through a component or device; let X be the real-number coefficient of the reactance (in ohms) of the component or device. Then the AC voltage (in volts rms), V_{rms}, across the component or device is given by:

$$V_{rms} = |I_{rms} \times X|$$

Alternating Current   169

Voltage vs. current, frequency, and inductance

Let I_{rms} be the alternating current (in amperes rms) through a component or device; let f be the AC frequency (in hertz); let L be the inductance (in henrys) of the component or device. Then the AC voltage (in volts rms), V_{rms}, across the component or device is given by:

$$V_{rms} = 2 \times \pi \times I_{rms} \times f \times L \approx 6.28 \times I_{rms} \times f \times L$$

This formula also applies for f in kilohertz (kHz) and L in millihenrys (mH); for f in megahertz (MHz) and L in microhenrys (µH); for f in gigahertz (GHz) and L in nanohenrys (nH); and for f in terahertz (THz) and L in picohenrys (pH).

Voltage vs. current, frequency, and capacitance

Let I_{rms} be the alternating current (in amperes rms) through a component or device; let f be the AC frequency (in hertz); let C be the capacitance (in farads) of the component or device. Then the AC voltage (in volts rms), V_{rms}, across the component or device is given by:

$$V_{rms} = I_{rms}/(2 \times \pi \times f \times C) \approx I_{rms}/(6.28 \times f \times C)$$

This formula also applies for f in megahertz (MHz) and C in microfarads (µF), and for f in terahertz (THz) and C in picofarads (pF).

Voltage vs. current and complex impedance

Let I_{rms} be the alternating current (in amperes rms) through a component or device; let the complex impedance of the component or device be $Z = R + jX$, where X is the real-number coefficient of the reactance (in ohms) of the component or device and R is the resistance (in ohms) of the component or device. Then the AC voltage (in volts rms), V_{rms}, is given by:

$$V_{rms} = I_{rms} \times (R^2 + X^2)^{1/2}$$

AC Power

The three ways of expressing AC power are as *real power* (in watts rms), as *reactive power* (in watts reactive), and as *apparent power* (in volt-amperes).

Real power

Let V_{rms} be the AC voltage across a component or device (in volts rms); let I_{rms} be the alternating current through the component or device (in amperes rms); let R be the resistance of the component or device (in ohms); let ϕ be the phase angle between the voltage and current waves. Then the real power (P_R) dissipated or radiated by the component or device (in watts rms) is given by the following formulas:

$$P_R = V_{rms} \times I_{rms} \times \cos \phi$$
$$P_R = I_{rms}^2 \times R \times \cos \phi$$
$$P_R = (V_{rms}^2/R) \times \cos \phi$$

Reactive power

Let V_{rms} be the AC voltage across a component or device (in volts rms); let I_{rms} be the alternating current through the component or device (in amperes rms); let |X| be the absolute value of the real-number coefficient of the reactance of the component or device (in ohms); let ϕ be the phase angle between the voltage and current waves. Then the reactive power (P_X) manifested in the component or device (in watts reactive) is given by the following formulas:

$$P_X = I_{rms}^2 \times |X|$$
$$P_X = V_{rms}^2/|X|$$
$$P_X = V_{rms} \times I_{rms} \times \sin \phi$$

Apparent power

Let V_{rms} be the AC voltage across a component or device (in volts rms); let I_{rms} be the alternating current through the component or device (in amperes rms); let P_R be the real power dissipated or radiated by the component or device (in watts rms); let P_X be the reactive power manifested in the component or device (in watts reactive). Then the apparent power (P_{VA}), dissipated or radiated by the component or device (in volt-amperes) is given by the following formulas:

$$P_{VA} = V_{rms} \times I_{rms}$$
$$P_{VA} = (P_R^2 + P_X^2)^{1/2}$$

AC Energy

The three ways of expressing AC energy are as *real energy* (in joules), as *reactive energy* (in joules reactive), or as *apparent energy* (in volt-ampere-seconds).

Real energy

Let V_{rms} be the AC voltage across a component or device (in volts rms); let I_{rms} be the alternating current through the component or device (in amperes rms); let R be the resistance of the component or device (in ohms); let ϕ be the phase angle between the voltage and current waves. Then the real energy, E_R, dissipated or radiated by the component or device (in joules) over period of time t (in seconds) is given by the following formulas:

$$E_R = V_{rms} \times I_{rms} \times t \times \cos\phi$$
$$E_R = I_{rms}^2 \times R \times t \times \cos\phi$$
$$E_R = (V_{rms}^2/R) \times t \times \cos\phi$$

Reactive energy

Let V_{rms} be the AC voltage across a component or device (in volts rms); let I_{rms} be the alternating current through the component or device (in amperes rms); let |X| be the absolute value of the real-number coefficient of the reactance of the component or device (in ohms); let ϕ be the phase angle between the voltage and current waves. Then the reactive energy, E_X, manifested in the component or device (in joules reactive) over period of time t (in seconds) is given by the following formulas:

$$E_X = I_{rms}^2 \times t \times |X|$$
$$E_X = (V_{rms}^2 \times t)/|X|$$
$$E_X = V_{rms} \times I_{rms} \times t \times \sin\phi$$

Apparent energy

Let V_{rms} be the AC voltage across a component or device (in volts rms); let I_{rms} be the alternating current through the component or device (in amperes rms); let E_R be the real energy dissipated or radiated by the component or device (in joules); let E_X be the reactive energy manifested in the component or device (in joules reactive). Then the apparent energy, E_{VA}, dissipated or radiated by the component or device (in volt-ampere-seconds) over period of time t (in seconds) is given by the following formulas:

$$E_{VA} = V_{rms} \times I_{rms} \times t$$
$$E_{VA} = (E_R^2 + E_X^2)^{1/2}$$

Chapter 10
Magnetism and Transformers

This chapter contains formulas involving magnetic fields, magnetic circuits, and transformer behavior.

Reluctance

Reluctance is a measure of the opposition that a circuit offers to the establishment of a magnetic field. It is symbolized R and is measured in *ampere-turns per weber* in the SI system of units.

Reluctance of a magnetic core

Let s represent the length (in meters) of a path through a magnetic core; let μ represent the magnetic permeability of the core material (in teslameters per ampere); let A represent the cross-sectional area of the core (in square meters). Then the reluctance, R (in ampere-turns per weber), is:

$$R = s/(\mu \times A)$$

This formula also holds for s in centimeters, μ in gauss per oersted, and A in square centimeters.

Reluctances in series

Reluctances in series add like resistances in series. If $R_1, R_2, R_3, \ldots R_n$ represent reluctances and R_s represents their series combination, then:

$$R_s = R_1 + R_2 + R_3 + \ldots + R_n$$

Reluctances in parallel

Reluctances in parallel add like resistances in parallel. If $R_1, R_2, R_3, \ldots R_n$ represent reluctances and R_p represents their parallel combination, then:

$$R_p = 1/(1/R_1 + 1/R_2 + 1/R_3 + \ldots + 1/R_n)$$

For only two reluctances (R_1 and R_2) in parallel, the composite reluctance, R_p, is given by:

$$R_p = (R_1 \times R_2) / (R_1 + R_2)$$

Basic Formulas

The following formulas outline the properties of simple magnetic circuits in which either of the following occur:

- An electrical conductor carries current that produces a magnetic field around the conductor
- A conductor moves relative to magnetic lines of flux, inducing current in the conductor

Flux density

Let Φ represent magnetic flux (in webers); let A represent the cross-sectional area of a region through which the flux lines pass at right angles (in square meters). Then the magnetic flux density, B (in teslas), is given by:

$$B = \Phi/A$$

This formula also holds when B is specified in gauss, Φ is specified in maxwells, and A is specified in square centimeters.

Permeability

Let B represent magnetic flux density (in teslas); let H represent magnetic field intensity (in amperes per meter). Then the permeability, μ (in tesla-meters per ampere), is given by:

$$\mu = B/H$$

This formula also holds when μ is specified in gauss per oersted, B is in gauss, and H is in oersteds.

Magnetomotive force

Let N represent the number of turns in an air-core coil; let I represent the current through the coil (in amperes). Then the magnetomotive force, F (in ampere-turns), is given by:

$$F = N \times I$$

If F is specified in gilberts, then:

$$F = 0.4 \times \pi \times N \times I \approx 1.256 \times N \times I$$

Magnetizing force

Let N represent the number of turns in an air-core coil; let I represent the current through the coil (in amperes); let s represent the length of a magnetic path through the coil (in meters). Then the magnetizing force, H (in ampere-turns per meter), is:

$$H = N \times I/s$$

If H is specified in oersteds and s is specified in centimeters, then:

$$H = 0.4 \times \pi \times N \times I/s \approx 1.256 \times N \times I/s$$

Induced Voltage

When a conductor moves relative to a magnetic field, a voltage is generated between the ends of the conductor. In practical circuits, this motion occurs in either of the following two ways:

- The conductor moves across stationary, constant magnetic lines of flux
- The magnetic field varies in magnitude around a stationary coil

Conductor in motion

Let B represent the intensity of a stationary, constant magnetic field (in webers per square meter); let s represent the length of a conductor (in meters); let v represent the velocity of the conductor (in meters per second) at right angles to the magnetic lines of flux. Then the induced voltage,

V (in volts), between the ends of the conductor is given by:

$$V = B \times s \times v$$

Variable flux

Let N represent the number of turns in a coil; let $d\Phi/dt$ represent the change in magnetic flux (in webers per second). Then the induced voltage (V) between the ends of the coil is given by:

$$V = N \times (d\Phi/dt)$$

Transformers

Transformers are generally used in electrical and electronic systems for either of two purposes:

- To step an AC voltage up or down
- To match impedances between two AC circuits or stages

Transformer efficiency

Let I_{pri} represent the current (in amperes) in the primary winding of a transformer; let I_{sec} represent the current (in amperes) in the secondary; let V_{pri} represent the rms sine-wave AC voltage across the primary; let V_{sec} represent the rms sine-wave AC voltage across the secondary. Then the efficiency (*Eff*) of the transformer is given (as a ratio) by:

$$Eff = (V_{sec} \times I_{sec}) / (V_{pri} \times I_{pri})$$

Expressed as a percentage and denoted *Eff%*, the efficiency of the transformer is:

$$Eff\% = (100 \times V_{sec} \times I_{sec})/(V_{pri} \times I_{pri})$$

P:S turns ratio

Let N_{pri} represent the number of turns in the primary winding of a transformer; let N_{sec} represent the number of turns in the secondary winding. Then the primary-to-secondary (*P:S*) turns ratio of the transformer is:

$$P{:}S = N_{pri} / N_{sec}$$

S:P turns ratio

The secondary-to-primary (*S:P*) turns ratio of a transformer is given by:

$$S{:}P = N_{sec}/N_{pri} = 1/(P{:}S)$$

Voltage transformation

Let V_{pri} represent the rms sine-wave AC voltage across the primary winding of a transformer (in volts). Then the rms sine-wave AC voltage across the secondary, V_{sec} (in volts), neglecting transformer losses, is given by:

$$V_{\text{sec}} = (S{:}P) \times V_{\text{pri}}$$

Impedance transformation

Let $S{:}P$ represent the secondary-to-primary turns ratio of a transformer; let $Z_{\text{in}} = R_{\text{in}} + j0$ represent a purely resistive (zero-reactance) impedance at the input (across the primary winding). Then the impedance at the output (across the secondary winding), Z_{out}, is also purely resistive, and is given by:

$$Z_{\text{out}} = (S{:}P)^2 \times Z_{\text{in}} = (S{:}P)^2 \times R_{\text{in}} + j0$$

Let $P{:}S$ represent the primary-to-secondary turns ratio of a transformer; let $Z_{\text{sec}} = R_{\text{sec}} + j0$ represent a purely resistive (zero-reactance) impedance connected across the secondary winding. Then the reflected impedance across the primary winding, Z_{pri}, is also purely resistive, and is given by:

$$Z_{\text{pri}} = (P{:}S)^2 \times Z_{\text{sec}} = (P{:}S)^2 \times R_{\text{sec}} + j0$$

Current demand

Let I_{load} represent the rms sine-wave alternating current drawn by a load connected to the secondary winding of a transformer (in amperes). Then the rms sine-wave alternating current demanded from a power source connected to the primary, I_{src} (in amperes), neglecting transformer losses, is given by:

$$I_{src} = (S{:}P) \times I_{load}$$

Losses in Transformers and Inductors

Losses occur in transformers and conductors as a result of conductor resistance and the properties of the core material.

Ohmic loss

Let I_{rms} represent the alternating current (in amperes rms) through an inductor or transformer winding; let V_{rms} represent the AC voltage (in volts rms) across the inductor or winding; let R represent the resistive component of the complex impedance of the inductor or winding (in ohms). Then the ohmic loss, P_Ω (in watts), is given by either of the following two formulas:

$$P_\Omega = I_{rms}^2 \times R$$
$$P_\Omega = V_{rms}^2 / R$$

Eddy-current loss

Let B represent the maximum flux density in an inductor or transformer core (in gauss); let s represent the thickness of the core material (in centimeters); let U represent the volume of the core material (in cubic centimeters); let f represent the frequency of the applied alternating current (in hertz); let k represent the core constant as specified by the manufacturer. Then the eddy-current loss, P_I (in watts), is given by:

$$P_I = k \times U \times B \times s^2 \times f^2$$

At 60 Hz, the utility line frequency commonly used in the United States:

$$P_I = 3.6 \times 10^3 \times k \times U \times B \times s^2$$

In circuits where the AC line frequency is 50 Hz:

$$P_I = 2.5 \times 10^3 \times k \times U \times B \times s^2$$

For silicon steel, a core material used in AC power-supply transformers, the core constant (k) is typically in the neighborhood of 4×10^{-12}. However, the value of k can differ substantially from this in the case of powdered-iron cores.

Hysteresis loss

Let $A_{\text{B-H}}$ represent the area of the measured hysteresis curve (B-H curve) for a core material at a specific frequency, where B is the flux density in gauss and H is the magnetizing force in oersteds. Then the hysteresis loss, P_H (in watts), is given by:

$$P_H = 0.796 \times 10^{-8} \times A_{B\text{-}H}$$

Total loss in transformer

Let V_{pri} represent the AC voltage (in volts rms) across the primary winding of a transformer operating with a specific constant load. Let V_{sec} represent the AC voltage (in volts rms) across the secondary; let I_{pri} represent the alternating current (in amperes rms) through the primary; let I_{sec} represent the alternating current (in amperes rms) through the secondary. Then, assuming zero reactance in the source or the load, the total power loss in the transformer, P_L (in watts), is given by:

$$P_L = V_{pri} \times I_{pri} - V_{sec} \times I_{sec}$$

Total loss in inductor or winding

Let P_Ω represent the ohmic loss (in watts) in an inductor or transformer winding; let P_I represent the eddy-current loss (in watts); let P_H represent the hysteresis loss (in watts); let P_Φ represent flux-leakage loss (in watts). Then the total power loss, P_L (in watts), is given by:

$$P_L = P_\Omega + P_I + P_H + P_\Phi$$

Chapter 11

Digital Electronics

This chapter contains formulas and tables relating to numbering systems, binary logic, and Boolean algebra (also known as *sentential logic*).

Numbering Systems

Four numbering systems are commonly used in electronics. The *decimal system* (radix or base 10) is used in conventional analog calculations. *Binary*, *octal*, and *hexadecimal systems* (radix or base 2, 8, and 16, respectively) are used in digital circuits, including computers.

Decimal (radix 10)

Digits in radix-10 numerals are from the following set:

$$N = \{0, 1, 2, 3, 4, 5, 6, 7, 8, 9\}$$

Values to the left of the radix point increase in powers of 10; values to the right of the radix point decrease in powers of 10. Let i be a decimal integer subscript denoting the physical position of a digit in a decimal numeral. If n_i represent the digits where each $n_i \in N$, then:

$$\ldots n_2\, n_1\, n_0\, .\, n_{-1}\, n_{-2} \ldots$$
$$= \ldots + n_2 \times 10^2 + n_1 \times 10 + n_0 + n_{-1} \times 10^{-1} + n_{-2} \times 10^{-2} + \ldots$$

Binary numbers (radix 2)

Digits in radix-2 numerals are from the following set:

$$N = \{0, 1\}$$

Values to the left of the radix point increase in powers of 2; values to the right of the radix point decrease in powers of 2. Let i be a decimal integer subscript denoting the physical position of a digit in a binary numeral. If n_i represent the digits where each $n_i \in N$, then:

$$\ldots n_2\, n_1\, n_0\, .\, n_{-1}\, n_{-2} \ldots$$
$$= \ldots + n_2 \times 2^2 + n_1 \times 2 + n_0 + n_{-1} \times 2^{-1} + n_{-2} \times 2^{-2} + \ldots$$

Octal numbers (radix 8)

Digits in radix-8 numerals are from the following set:

$$N = \{0, 1, 2, 3, 4, 5, 6, 7\}$$

Values to the left of the radix point increase in powers of 8; values to the right of the radix point decrease in powers of 8. Let i be a decimal integer subscript denoting the physical position of a digit in an octal numeral. If n_i represent the digits where each $n_i \in N$, then:

$$\ldots n_2\, n_1\, n_0\, .\, n_{-1}\, n_{-2} \ldots$$
$$= \ldots + n_2 \times 8^2 + n_1 \times 8 + n_0 + n_{-1} \times 8^{-1} + n_{-2} \times 8^{-2} + \ldots$$

Hexadecimal numbers (radix 16)

Digits in radix-16 numerals are from the following set:

N = {0, 1, 2, 3, 4, 5, 6, 7, 8, 9, A, B, C, D, E, F}

Values to the left of the radix point increase in powers of 16; values to the right of the radix point decrease in powers of 16. Let i be a decimal integer subscript denoting the physical position of a digit in a hexadecimal numeral. If n_i represent the digits where each $n_i \in N$, then:

$$\ldots n_2 \, n_1 \, n_0 \, . \, n_{-1} \, n_{-2} \ldots$$
$$= \ldots + n_2 \times 16^2 + n_1 \times 16 + n_0$$
$$+ n_{-1} \times 16^{-1} + n_{-2} \times 16^{-2} + \ldots$$

Number conversion

Table 11.1 compares decimal, binary, octal, and hexadecimal numbers for decimal values 0 through 64.

TABLE 11.1 Comparison of Values in Decimal, Binary, Octal, and Hexadecimal Numbering Systems For Decimal 0 Through 64

Decimal	Binary	Octal	Hexadecimal
0	0	0	0
1	1	1	1
2	10	2	2
3	11	3	3

TABLE 11.1 *(Continued)*

Decimal	Binary	Octal	Hexadecimal
4	100	4	4
5	101	5	5
6	110	6	6
7	111	7	7
8	1000	10	8
9	1001	11	9
10	1010	12	A
11	1011	13	B
12	1100	14	C
13	1101	15	D
14	1110	16	E
15	1111	17	F
16	10000	20	10
17	10001	21	11
18	10010	22	12
19	10011	23	13
20	10100	24	14
21	10101	25	15
22	10110	26	16
23	10111	27	17
24	11000	30	18
25	11001	31	19
26	11010	32	1A

Chapter Eleven

TABLE 11.1 *(Continued)*

Decimal	Binary	Octal	Hexadecimal
27	11011	33	1B
28	11100	34	1C
29	11101	35	1D
30	11110	36	1E
31	11111	37	1F
32	100000	40	20
33	100001	41	21
34	100010	42	22
35	100011	43	23
36	100100	44	24
37	100101	45	25
38	100110	46	26
39	100111	47	27
40	101000	50	28
41	101001	51	29
42	101010	52	2A
43	101011	53	2B
44	101100	54	2C
45	101101	55	2D
46	101110	56	2E
47	101111	57	2F
48	110000	60	30
49	110001	61	31

TABLE 11.1 *(Continued)*

Decimal	Binary	Octal	Hexadecimal
50	110010	62	32
51	110011	63	33
52	110100	64	34
53	110101	65	35
54	110110	66	36
55	110111	67	37
56	111000	70	38
57	111001	71	39
58	111010	72	3A
59	111011	73	3B
60	111100	74	3C
61	111101	75	3D
62	111110	76	3E
63	111111	77	3F
64	1000000	80	40

Basic Binary Operations

The basic binary operations are *negation* (NOT), *conjunction* (AND), and *disjunction* (OR). The two possible values for variables in these operations are logic 0 (low) and logic 1 (high). Variables

NOT operation

NOT X is written $-X$, as X with a tilde or line over it, or as X'. Given the logic equation $-X = Z$, the output $Z = 1$ if $X = 0$; the output $Z = 0$ if $X = 1$ (see Table 11.2).

AND operation

X AND Y is written $X \times Y$, XY, or $X * Y$. Given the logic equation $X \times Y = Z$, the output is $Z = 1$ if and only if $X = 1$ and $Y = 1$; otherwise, the output is $Z = 0$ (see Table 11.2).

TABLE 11.2 Basic Binary Operations

X	Y	$-X$	$X \times Y$	$X + Y$
0	0	1	0	0
0	1	1	0	1
1	0	0	0	1
1	1	0	1	1

OR operation

X OR Y is written $X + Y$. Given $X + Y = Z$, the output is $Z = 1$ if $X = 1$, if $Y = 1$, or if $X = 1$ and $Y = 1$. If $X = 0$ and $Y = 0$, then the output is $Z = 0$ (see Table 11.2).

Secondary Binary Operations

The secondary binary operations are NOT-AND or NAND, not-OR or NOR, and exclusive-OR or XOR.

NAND operation

Given the logic equation X NAND $Y = Z$, the output is $Z = 0$ if and only if $X = 1$ and $Y = 1$; otherwise, the output $Z = 1$ (see Table 11.3).

TABLE 11.3 Secondary Binary Operations

X	Y	X NAND Y	X NOR Y	X XOR Y
0	0	1	1	0
0	1	1	0	1
1	0	1	0	1
1	1	0	0	0

NOR operation

Given the logic equation X NOR $Y = Z$, the output is $Z = 1$ if and only if $X = 0$ and $Y = 0$; otherwise, the output is $Z = 0$ (see Table 11.3).

XOR operation

Given the logic equation X XOR $Y = Z$, the output $Z = 0$ if and only if $X = Y$; the output $Z = 1$ if and only if $X \neq Y$ (see Table 11.3).

Logic Gates

Logic gates are electrical switches that perform binary logic functions. Most gates operate with logic 0 (low) represented by a signal of about 0 volts DC, and logic 1 (high) represented by a signal of about +5 volts DC. Schematic symbols for logic gates are shown in Figure 11.1.

Inverter (NOT gate)

An *inverter* or *NOT gate* has one input and one output. It reverses the state of the input signal (see Table 11.4).

AND gate

An *AND gate* can have two or more inputs. If all the input signals are high, then the output signal is

Digital Electronics 197

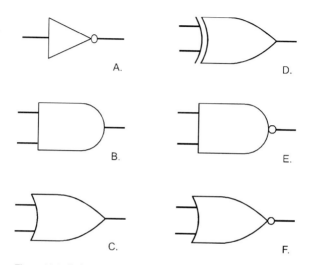

Figure 11.1 An inverter or NOT gate (A), an AND gate (B), an OR gate (C), an XOR gate (D), a NAND gate (E), and a NOR gate (F).

TABLE 11.4 Logic Gates and Their Characteristics.

Gate type	Number of inputs	Remarks
NOT	1	Changes state of input.
AND	2 or more	Output low if any inputs are low.
		Output high if all inputs are high.
OR	2 or more	Output high if any inputs are high.

TABLE 11.4 *(Continued)*

Gate type	Number of inputs	Remarks
NAND	2 or more	Output low if all inputs are low.
		Output high if any inputs are low.
		Output low if all inputs are high.
NOR	2 or more	Output low if any inputs are high.
		Output high if all inputs are low.
XOR	2	Output high if inputs differ.
		Output low if inputs are the same.

high. Otherwise, the output signal is low (see Table 11.4).

OR gate

An *OR gate* can have two or more inputs. If all the input signals are low, then the output signal is low. Otherwise, the output signal is high (see Table 11.4).

NAND gate

If an AND gate is followed by an inverter, the result is a *NAND gate*. If all of the input signals are high, then the output signal is low. Otherwise, the output signal is high (see Table 11.4).

NOR gate

If an OR gate is followed by an inverter, the result is a *NOR gate*. If all the input signals are low, then the output signal is high. Otherwise, the output signal is low (see Table 11.4).

XOR gate

An *exclusive OR gate*, also called an *XOR gate*, has two inputs and one output. If the input signals are the same, then the output signal is low. If the input signals are different, then the output signal is high.

Boolean Theorems

Table 11.5 shows several logic equations. These are facts, or theorems. Boolean theorems can be used to analyze complicated logic functions.

TABLE 11.5 Theorems in Boolean Algebra

Equation	Name (if applicable)
$X + 0 = X$	OR identity
$X \times 1 = X$	AND identity
$X + 1 = 1$	
$X \times 0 = 0$	
$X + X = X$	
$X \times X = X$	
$-(-X) = X$	Double negation
$X + (-X) = X$	
$X \times (-X) = 0$	Contradiction
$X + Y = Y + X$	Commutativity of OR
$X \times Y = Y \times X$	Commutativity of AND
$X + (X \times Y) = X$	
$X \times (-Y) + Y = X + Y$	
$X + Y + Z = (X + Y) + Z = X + (Y + Z)$	Associativity of OR
$X \times Y \times Z = (X \times Y) \times Z = X \times (Y \times Z)$	Associativity of AND
$X \times (Y + Z) = (X \times Y) + (X \times Z)$	Distributivity
$-(X + Y) = (-X) \times (-Y)$	DeMorgan's Theorem
$-(X \times Y) = (-X) + (-Y)$	DeMorgan's Theorem

Flip-flops

A *flip-flop* is a form of sequential logic gate. In a sequential gate, the output state depends on both the inputs and the outputs. A flip-flop has two states, called *set* and *reset*. Usually, the set state is logic 1 (high), and the reset state is logic 0 (low).

R-S flip-flop

R-S flip-flop inputs are labeled R (reset) and S (set). The outputs are labeled Q and $-Q$. The symbol for an R-S flip-flop, also known as an *asynchronous flip-flop*, is shown at Figure 11.2A. The truth table for an R-S flip-flop is in Table 11.6.

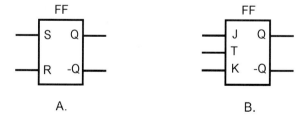

Figure 11.2 Symbol for an R-S flip-flop (A); symbol for a J-K flip-flop (B).

TABLE 11.6 R-S Flip-Flop States

R	S	Q	–Q
0	0	Q	–Q
0	1	1	0
1	0	0	1
1	1	?	?

M-S flip-flop

A *master-slave* (*M-S*) *flip-flop* essentially consists of two R-S flip-flops in series. The first flip-flop is the *master*, and the second is the *slave*. The master functions when the clock output is high, and the slave acts during the next low portion of the clock output. This delay prevents confusion between the input and output.

J-K flip-flop

J-K flip-flop operation is similar to that of an R-S flip-flop, except that the J-K has a predictable output when the inputs are both 1. Table 11.7 shows the input and output states for this type of flip-flop. The output changes only when a triggering

TABLE 11.7 J-K Flip-Flop States

R	S	Q	–Q
0	0	Q	–Q
0	1	1	0
1	0	0	1
1	1	–Q	Q

pulse is received. The symbol for a J-K flip-flop is shown in Figure 11.2C.

R-S-T flip-flop

R-S-T flip-flop operation is similar to that of an R-S flip-flop, except that a high pulse at the T input causes the circuit to change state.

T flip-flop

The *T flip-flop* has only one input. Each time that a high pulse appears at the T input, the output state is reversed.

Chapter 12

Resonance, Filters, and Noise

This chapter contains formulas relevant to resonance, filter design, and noise characteristics.

Resonant Frequency

The *resonant frequency* is an important property of filters, antennas, and various other electronic devices. The following formulas concern RF electrical resonance.

Basic LC circuit

Let L be the inductance (in henrys) and C be the capacitance (in farads) in an inductance-capacitance (LC) resonant circuit. Then the resonant frequency f (in hertz) is given by:

$$f = 1/(2 \times \pi \times L^{1/2} \times C^{1/2})$$

This formula also holds for f in megahertz, L in microhenrys, and C in microfarads.

Air cavity (1/4 wave)

Let s be the end-to-end length (in inches) of an air cavity. Then the fundamental quarter-wave resonant frequency, f (in megahertz), is given by:

$$f = 2.95 \times 10^3/s$$

If s is in centimeters, then:

$$f = 7.50 \times 10^3/s$$

Harmonic quarter-wave resonances occur at odd integral multiples of this frequency.

Air cavity (1/2 wave)

Let s be the end-to-end length (in inches) of an air cavity. Then the fundamental half-wave resonant frequency, f (in megahertz), is given by:

$$f = 4.90 \times 10^3 / s$$

If s is in centimeters, then:

$$f = 1.50 \times 10^4 / s$$

Harmonic half-wave resonances occur at all integral multiples of this frequency.

Transmission-line section (1/4-wave)

Let s be the end-to-end length (in inches) of a section of transmission line whose velocity factor (as a ratio between 0 and 1) is v. Then the fundamental quarter-wave resonant frequency, f (in megahertz), is given by:

$$f = 2.95 \times 10^3 \times v / s$$

If s is in centimeters, then:

$$f = 7.50 \times 10^3 \times v / s$$

If s is in feet, then:

$$f = 246 \times v/s$$

Harmonic quarter-wave resonances occur at odd integral multiples of this frequency.

Transmission-line section (1/2-wave)

Let s be the end-to-end length (in inches) of a section of transmission line whose velocity factor (as a ratio between 0 and 1) is v. Then the fundamental half-wave resonant frequency, f (in megahertz), is given by:

$$f = 4.90 \times 10^3 \times v/s$$

If s is in centimeters, then:

$$f = 1.50 \times 10^4 \times v/s$$

If s is in feet, then:

$$f = 492 \times v/s$$

Harmonic half-wave resonances occur at all integral multiples of this frequency.

Lowpass Filters

A *lowpass filter* offers little or no attenuation of signals at frequencies less than the *cutoff*, and significant attenuation of signals at frequencies greater than the cutoff.

Constant-*k*

Let f be the cutoff frequency (in hertz) for a constant-k lowpass LC filter, as shown in Figure 12.1.

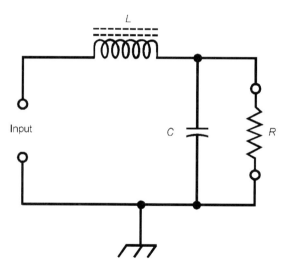

Figure 12.1 Constant-k lowpass filter.

Let R be the load resistance (in ohms). Then the optimum inductance, L (in henrys), is given by:

$$L = R/(\pi \times f)$$

The optimum capacitance, C (in farads), for the filter shown in Figure 12.1 is given by:

$$C = 1/(\pi \times f \times R)$$

Series *m*-derived

Let f_1 be the highest frequency of maximum transmission (in hertz) for a series m-derived lowpass LC filter, as shown in Figure 12.2. Let f_2 be the lowest frequency of maximum attenuation (in hertz); let R be the load resistance (in ohms). Then the filter constant m is given by:

$$m = (1 - f_1^2/f_2^2)^{1/2}$$

The optimum inductance, L_1 (in henrys), for the filter shown in Figure 12.2 is given by:

$$L_1 = m \times R/(\pi \times f_1)$$

The optimum inductance, L_2 (in henrys), for the filter shown in Figure 12.2 is given by:

$$L_2 = R \times (1 - m^2)/(4 \times \pi \times m \times f_1)$$

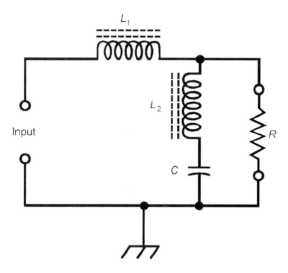

Figure 12.2 Series m-derived lowpass filter.

The optimum capacitance, C (in farads), for the filter shown in Figure 12.2 is given by:

$$C = (1 - m^2)/(\pi \times f_1)$$

Shunt *m*-derived

Let f_1 be the highest frequency of maximum transmission (in hertz) for a shunt m-derived lowpass

LC filter, as shown in Figure 12.3. Let f_2 be the lowest frequency of maximum attenuation (in hertz); let R be the load resistance (in ohms). Then the filter constant (m) is given by:

$$m = (1 - f_1^2/f_2^2)^{1/2}$$

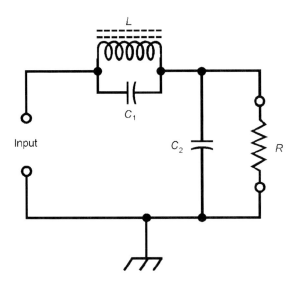

Figure 12.3 Shunt m-derived lowpass filter.

The optimum inductance, L (in henrys), for the filter shown in Figure 12.3 is given by:

$$L = m \times R/(\pi \times f_1)$$

The optimum capacitance, C_1 (in farads), for the filter shown in Figure 12.3 is given by:

$$C_1 = (1 - m^2)/(4 \times \pi \times R \times m \times f_1)$$

The optimum capacitance, C_2 (in farads), for the filter shown in Figure 12.3 is given by:

$$C_2 = m/(\pi \times R \times f_2)$$

Highpass Filters

A *highpass filter* offers little or no attenuation of signals at frequencies greater than the *cutoff*, and significant attenuation of signals at frequencies less than the cutoff.

Constant-*k*

Let f be the cutoff frequency (in hertz) for a constant-k highpass LC filter, as shown in Figure 12.4. Let R be the load resistance (in ohms). Then the optimum inductance, L (in henrys), is given by:

$$L = R/(4 \times \pi \times f)$$

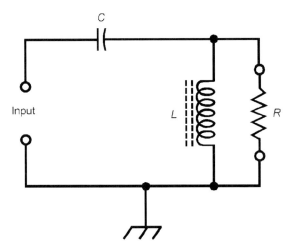

Figure 12.4 Constant-k highpass filter.

The optimum capacitance C (in farads) for the filter shown in Figure 12.4 is given by:

$$C = 1/(4 \times \pi \times f \times R)$$

Series *m*-derived

Let f_1 be the highest frequency of maximum attenuation (in hertz) for a series m-derived highpass LC filter as shown in Figure 12.5. Let f_2 be the

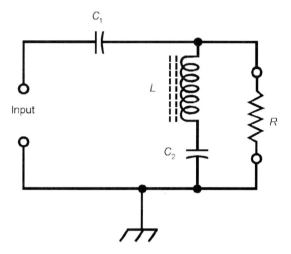

Figure 12.5 Series m-derived highpass filter.

lowest frequency of maximum transmission (in hertz); let R be the load resistance (in ohms). Then the filter constant (m) is given by:

$$m = (1 - f_1^2/f_2^2)^{1/2}$$

The optimum inductance, L (in henrys), for the filter shown in Figure 12.5 is given by:

$$L = R/(4 \times \pi \times m \times f_2)$$

The optimum capacitance, C_1 (in farads), for the filter shown in Figure 12.5 is given by:

$$C_1 = 1/(4 \times \pi \times m \times f_2 \times R)$$

The optimum capacitance, C_2 (in farads), for the filter shown in Figure 12.5 is given by:

$$C_2 = m/[(1 - m^2) \times \pi \times f_2 \times R]$$

Shunt *m*-Derived

Let f_1 be the highest frequency of maximum attenuation (in hertz) for a shunt m-derived high-pass LC filter, as shown in Figure 12.6. Let f_2 be the lowest frequency of maximum transmission (in hertz); let R be the load resistance (in ohms). Then the filter constant (m) is given by:

$$m = (1 - f_1^2/f_2^2)^{1/2}$$

The optimum inductance, L_1 (in henrys), for the filter shown in Figure 12.6 is given by:

$$L_1 = m \times R/[(1 - m^2) \times \pi \times f_2]$$

The optimum inductance, L_2 (in henrys), for the filter shown in Figure 12.6 is given by:

$$L_2 = R/(4 \times \pi \times m \times f_2)$$

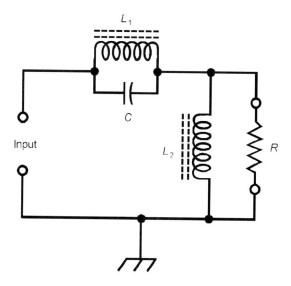

Figure 12.6 Shunt m-derived highpass filter.

The optimum capacitance, C (in farads), for the filter shown in Figure 12.6 is given by:

$$C = 1/(4 \times \pi \times m \times f_2 \times R)$$

Bandpass Filters

A *bandpass filter* offers little or no attenuation of signals whose frequencies are between a *lower*

Chapter Twelve

cutoff and an *upper cutoff*, and significant attenuation of signals whose frequencies are outside of this range.

Constant-*k*

Let f_1 be the lower cutoff frequency (in hertz) for a constant-*k* bandpass *LC* filter, as shown in Figure 12.7. Let f_2 be the upper cutoff frequency (in hertz). Let *R* be the load resistance (in ohms). Then the optimum inductances (in henrys) are given by:

$$L_1 = R/[\pi \times (f_2 - f_1)]$$

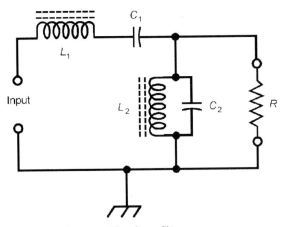

Figure 12.7 Constant-*k* bandpass filter.

$$L_2 = (f_2 - f_1) \times R / (4 \times \pi \times f_1 \times f_2)$$

The optimum capacitances (in farads) for the filter shown in Figure 12.7 are given by:

$$C_1 = (f_2 - f_1)/(4 \times \pi \times f_1 \times f_2 \times R)$$
$$C_2 = 1/[\pi \times (f_2 - f_1) \times R]$$

Series *m*-derived

Let frequencies $f_1, f_2, f_3,$ and f_4 be expressed in hertz and defined, as shown in Figure 12.8. Let R be the

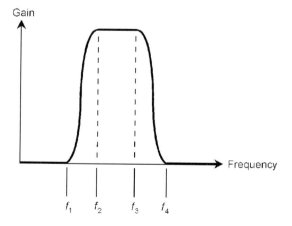

Figure 12.8 Bandpass filter response curve.

load resistance (in ohms). Define the quantities x, m (the filter constant), y, and z, as follows:

$$x = [(1 - f_2^2/f_3^2) \times (1 - f_3^2/f_4^2)]^{1/2}$$
$$m = x/(1 - f_2 \times f_3/f_4^2)$$
$$y = (1 - m^2) \times (1 - f_1^2/f_4^2) \times f_2 \times f_3/(4 \times x \times f_1^2)$$
$$z = (1 - m^2) \times (1 - f_1^2/f_4^2)/(4 \times x)$$

The optimum inductances (in henrys) for the filter shown in Figure 12.9 are given by:

$$L_1 = m \times R/[\pi \times (f_3 - f_2)]$$
$$L_2 = z \times R/[\pi \times (f_3 - f_2)]$$
$$L_3 = y \times R/[\pi \times (f_3 - f_2)]$$

The optimum capacitances (in farads) for the filter shown in Figure 12.9 are given by:

$$C_1 = (f_3 - f_2)/(4 \times \pi \times m \times R \times f_2 \times f_3)$$
$$C_2 = (f_3 - f_2)/(4 \times \pi \times y \times R \times f_2 \times f_3)$$
$$C_3 = (f_3 - f_2)/(4 \times \pi \times z \times R \times f_2 \times f_3)$$

Shunt *m*-derived

Let frequencies $f_1, f_2, f_3,$ and f_4 be expressed in hertz and defined, as shown in Figure 12.8. Let R be the load resistance (in ohms). Define the quantities x, m (the filter constant), y, and z, as in the preceding section for the series *m*-derived bandpass filter.

Resonance, Filters, and Noise 221

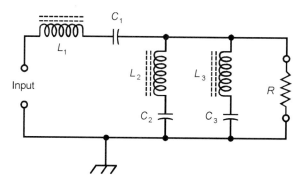

Figure 12.9 Series m-derived bandpass filter.

Then the optimum inductances (in henrys) for the filter shown in Figure 12.10 are given by:

$$L_1 = (f_3 - f_2) \times R/(4 \times \pi \times z \times f_2 \times f_3)$$
$$L_2 = (f_3 - f_2) \times R/(4 \times \pi \times y \times f_2 \times f_3)$$
$$L_3 = (f_3 - f_2) \times R/(4 \times \pi \times m \times f_2 \times f_3)$$

The optimum capacitances (in farads) for the filter shown in Figure 12.10 are given by:

$$C_1 = z/[\pi \times R \times (f_3 - f_2)]$$
$$C_2 = y/[\pi \times R \times (f_3 - f_2)]$$
$$C_3 = m/[\pi \times R \times (f_3 - f_2)]$$

222 Chapter Twelve

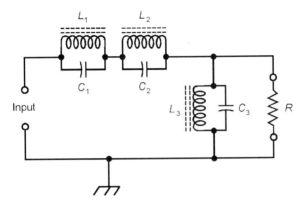

Figure 12.10 Shunt m-derived bandpass filter.

Bandstop Filters

A *bandstop filter*, also called a *band-rejection filter*, offers significant attenuation of signals whose frequencies are between a *lower cutoff* and an *upper cutoff*, and little or no attenuation of signals whose frequencies are outside this range.

Constant-*k*

Let f_1 be the lower cutoff frequency (in hertz) for a constant-k bandstop LC filter, as shown in

Figure 12.11. Let f_2 be the upper cutoff frequency (in hertz). Let R be the load resistance (in ohms). Then the optimum inductances (in henrys) are given by:

$$L_1 = R \times (f_2 - f_1)/(\pi \times f_1 \times f_2)$$
$$L_2 = R/[4 \times \pi \times (f_2 - f_1)]$$

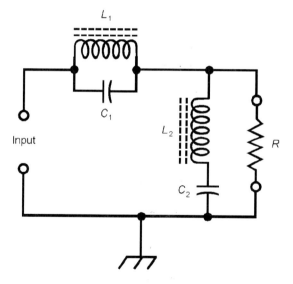

Figure 12.11 Constant-k bandstop filter.

The optimum capacitances (in farads) for the filter shown in Figure 12.11 are given by:

$$C_1 = 1/[(4 \times \pi \times R \times (f_2 - f_1)]$$
$$C_2 = (f_2 - f_1)/(\pi \times R \times f_1 \times f_2)$$

Series *m*-derived

Let frequencies f_1, f_2, f_3, and f_4 be expressed in hertz and defined, as shown in Figure 12.12. Let R

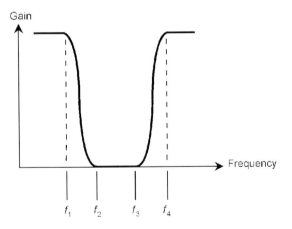

Figure 12.12 Bandstop filter response curve.

Resonance, Filters, and Noise

be the load resistance (in ohms). Define the quantities m (the filter constant), x, and y, as follows:

$$m = [(1 - f_1^2/f_3^2) \times (1 - f_3^2/f_4^2)/(1 - f_1/f_4)]^{1/2}$$
$$x = (1/m) \times (1 + f_1 \times f_4/f_3^2)$$
$$y = (1/m) \times [1 + f_3^2/(f_1 \times f_4)]$$

The optimum inductances (in henrys) for the filter shown in Figure 12.13 are given by:

$$L_1 = m \times R \times (f_4 - f_1)/(\pi \times f_1 \times f_4)$$
$$L_2 = R/[4 \times \pi \times m \times (f_4 - f_1)]$$
$$L_3 = y \times R/[4 \times \pi \times (f_4 - f_1)]$$

The optimum capacitances (in farads) for the filter shown in Figure 12.13 are given by:

$$C_1 = 1/[4 \times \pi \times m \times R \times (f_4 - f_1)]$$
$$C_2 = (f_4 - f_1)/(\pi \times y \times R \times f_1 \times f_4)$$
$$C_3 = (f_4 - f_1)/(\pi \times x \times R \times f_1 \times f_4)$$

Shunt *m*-derived

Let frequencies f_1, f_2, f_3, and f_4 be expressed in hertz and defined, as shown in Figure 12.12. Let R be the load resistance (in ohms). Define the quantities m (the filter constant), x, and y as in the

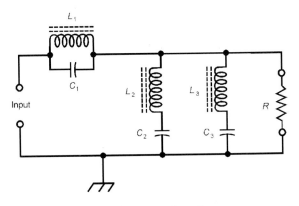

Figure 12.13 Series m-derived bandstop filter.

preceding section for the series m-derived bandstop filter. The optimum inductances (in henrys) for the filter shown in Figure 12.14 are given by:

$$L_1 = (f_4 - f_1) \times R / (\pi \times y \times f_1 \times f_4)$$
$$L_2 = (f_4 - f_1) \times R / (\pi \times x \times f_1 \times f_4)$$
$$L_3 = R / [4 \times \pi \times m \times (f_4 - f_1)]$$

The optimum capacitances (in farads) for the filter shown in Figure 12.14 are given by:

$$C_1 = x / [4 \times \pi \times R \times (f_4 - f_1)]$$
$$C_2 = y / [4 \times \pi \times R \times (f_4 - f_1)]$$
$$C_3 = m \times (f_4 - f_1) / (\pi \times R \times f_1 \times f_4)$$

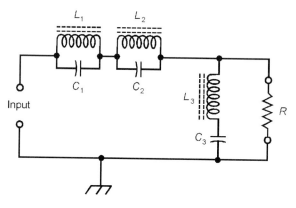

Figure 12.14 Shunt m-derived bandstop filter.

Noise

The following formulas deal with noise, and with common ways of expressing and measuring its practical effects.

Thermal noise power

Let k represent Boltzmann's constant (approximately 1.3807×10^{-23} joules per degree Kelvin); let T represent the absolute temperature (in degrees Kelvin); let B represent the bandwidth (in hertz). Then the thermal noise power (in watts), P_{nt}, is given by:

$$P_{nt} = k \times T \times B$$

Thermal noise voltage

Let R represent the resistance of a noise source (in ohms); let P_{nt} represent the thermal noise power (in watts). Then the thermal noise voltage (in volts), V_{nt}, is given by:

$$V_{nt} = (P_{nt} \times R)^{1/2}$$

Signal-to-noise ratio

Let P_n represent the noise power (in watts) at the output of a circuit; let P_s represent the signal power (in watts) at the output of the same circuit. Then the signal-to-noise ratio, $S{:}N$ (in decibels), is given by:

$$S{:}N = 10 \times \log_{10}(P_s/P_n)$$

The value of $S{:}N$ can also be calculated in terms of voltages or currents. Let V_n represent the noise voltage (in volts) at the output of a circuit; let I_n represent the noise current (in amperes) at that point; let V_s represent the signal voltage (in volts) at that point; let I_s represent the signal current (in amperes) at that point. Then the signal-to-noise ratio, $S{:}N$ (in decibels), assuming constant impedance, is given by either of these formulas:

$$S{:}N = 20 \times \log_{10} (V_s/V_n)$$
$$S{:}N = 20 \times \log_{10} (I_s/I_n)$$

Signal-plus-noise-to-noise ratio

Let P_n represent the noise power (in watts) at the output of a circuit; let P_s represent the signal power (in watts) at the output of the same circuit. Then the signal-plus-noise-to-noise ratio, *(S+N):N* (in decibels), is given by:

$$(S+N){:}N = 10 \times \log_{10} [(P_s + P_n)/P_n]$$

The value of *(S+N):N* can also be calculated in terms of voltages or currents. Let V_n represent the noise voltage (in volts) at the output of a circuit; let I_n represent the noise current (in amperes) at that point; let V_s represent the signal voltage (in volts) at that point; let I_s represent the signal current (in amperes) at that point. Then the signal-plus-noise-to-noise ratio, *(S+N):N* (in decibels), assuming constant impedance, is given by either of these formulas:

$$(S+N){:}N = 20 \times \log_{10} [(V_s + V_n)/V_n]$$
$$(S+N){:}N = 20 \times \log_{10} [(I_s + I_n)/I_n]$$

Noise figure

Let P_i represent the noise power (in watts) at the output of an ideal circuit; let P_a represent the noise power (in watts) at the output of an actual circuit. Then the noise figure, N (in decibels), of the actual circuit is given by:

$$N = 10 \times \log_{10} (P_a/P_i)$$

The noise figure can also be calculated in terms of $S{:}N$ ratios. Let $S{:}N_i$ be the $S{:}N$ ratio (in decibels) at the output of an ideal circuit; let $S{:}N_a$ be the $S{:}N$ ratio (in decibels) at the output of the actual circuit. Then the noise figure N (in decibels) of the actual circuit is given by:

$$N = 10 \times \log_{10} (S{:}N_i/S{:}N_a)$$

Chapter 13

Semiconductors

This chapter contains formulas involving semiconductor diodes, bipolar transistors, and field-effect transistors.

Diodes

A diode exhibits a nonlinear relationship between voltage and current. This relationship differs in the

Forward current

Let I_{rs} represent the reverse saturation current (in amperes) for a particular diode. Let q represent the charge on an electron (approximately 1.602×10^{-19} coulomb); let V_f represent the forward voltage (in volts); let k represent Boltzmann's constant (approximately 1.3807×10^{-23} joules per degree Kelvin); let T represent the absolute temperature (in degrees Kelvin); let e represent the exponential constant (approximately 2.718). Consider x to be defined as follows:

$$x = (q \times V_f)/(k \times T)$$

Then the *forward current*, I_f (in amperes), is given by:

$$I_f = I_{rs} \times (e^x - 1)$$

Static resistance

Let V_{DC} represent the DC voltage drop (in volts) across a diode; let I_{DC} represent the direct current (in amperes) through the diode. Then the *static resistance*, R_s (in ohms), of the diode is given by:

$$R_s = V_{DC}/I_{DC}$$

Dynamic resistance

Let V represent the instantaneous voltage drop (in volts) across a diode; let I represent the instantaneous current (in amperes) through the diode. Then the *dynamic resistance*, R_d (in ohms), of the diode is given by:

$$R_d = dV/dI$$

That is, R_d is the derivative of the voltage with respect to the current or the slope of the characteristic curve V versus I (Figure 13.1) at a specified point.

Rectification efficiency

Let V_{DC} represent the DC output voltage of a diode (in volts); let V_{pk} represent the peak AC input voltage (in volts). Then the *rectification efficiency*, η (in percent) is given by:

$$\eta = 100 \times V_{DC}/V_{pk}$$

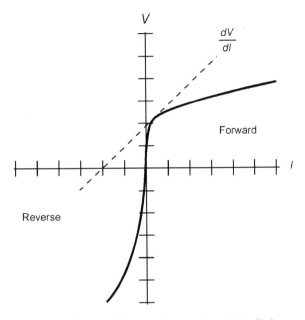

Figure 13.1 Characteristic curve for a semiconductor diode.

Bipolar Transistors

This section contains formulas relevant to bipolar transistors, both the NPN type and the PNP type.

Static forward current transfer ratio

Assume that the collector voltage, V_c, in a common-emitter circuit (Figure 13.2) is constant. Let

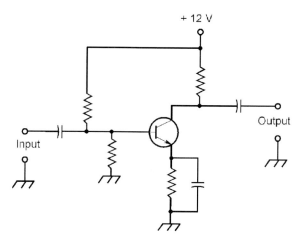

Figure 13.2 Common-emitter bipolar transistor circuit.

I_c represent the collector current (in amperes); let I_b represent the base current (in amperes). Then the *static forward current transfer ratio*, H_{FE}, is given by:

$$H_{FE} = I_c/I_b$$

Dynamic base resistance

Assume that the collector voltage, V_c, is constant. Let V_b represent the base voltage (in volts); let I_b represent the base current (in amperes). Then the *dynamic base resistance*, R_b (in ohms), is given by:

$$R_b = dV_b/dI_b$$

Dynamic emitter resistance

Assume that the collector voltage, V_c, applied to a transistor is constant. Let V_e represent the emitter voltage (in volts); let I_e represent the emitter current (in amperes). Then the *dynamic emitter resistance*, R_e (in ohms), is given by:

$$R_e = dV_e/dI_e$$

Dynamic collector resistance

Assume that the emitter current, I_e, through a transistor is constant. Let V_c represent the collector voltage (in volts); let I_c represent the collector current (in amperes). Then the *dynamic collector resistance*, R_c (in ohms), is given by:

$$R_c = dV_c/dI_c$$

Dynamic emitter feedback conductance

Assume that the emitter voltage, V_e, applied to a transistor is constant. Let I_e represent the emitter current (in amperes); let V_c represent the collector voltage (in volts). Then the *dynamic emitter feedback conductance*, G_{ec} (in siemens), is given by:

$$G_{ec} = dI_e/dV_c$$

Alpha

Assume that the collector voltage, V_c, in a transistor in a common-base circuit (Figure 13.3) is constant. Let I_c represent the collector current (in amperes); let I_e represent the emitter current (in amperes). Then the *dynamic current amplification in common-base arrangement*, or *alpha* (symbolized α), is given by:

$$\alpha = dI_c/dI_e$$

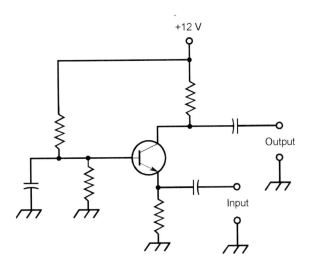

Figure 13.3 Common-base bipolar transistor circuit.

Beta

Assume that the collector voltage, V_c, applied to a transistor in a common-emitter circuit (Figure 13.2) is constant. Let I_c represent the collector current (in amperes); let I_b represent the base current (in amperes). Then the *dynamic current amplification in common-emitter arrangement*, or *beta* (symbolized β), is given by:

$$\beta = dI_c/dI_b$$

Alpha in terms of beta

Suppose that the beta (β) of a transistor is known. Then the alpha (α) of that transistor, assuming that the collector voltage (V_c) remains constant, is given by:

$$\alpha = \beta \, (1 + \beta)$$

Beta in terms of alpha

The beta (β), assuming that the collector voltage (V_c) remains constant, is given by:

$$\beta = \alpha/(1 - \alpha)$$

Dynamic stability factor

Let I_c represent the collector current through a bipolar transistor (in amperes); let I_ω represent the collector leakage current (in amperes). Then the *dynamic stability factor*, S, of the transistor is given by:

$$S = dI_c / dI_\omega$$

Resistance parameters (common base)

Let α represent the dynamic current amplification of a transistor in the common-base arrangement.

Let the following symbols represent resistances (in ohms):

R_b = dynamic base resistance
R_c = dynamic collector resistance
R_e = dynamic emitter resistance
R_{in} = input resistance
R_{rt} = reverse transfer resistance
R_{ft} = forward transfer resistance
R_{out} = output resistance

Then the following equations hold:

$$R_{in} = R_e + R_b$$
$$R_{rt} = R_b$$
$$R_{ft} = R_b + \alpha \times R_c$$
$$R_{out} = R_c + R_b$$

Resistance parameters (common emitter)

Let α represent the dynamic current amplification of a bipolar transistor in the common-base arrangement. Let the following symbols represent resistances (in ohms) in a common-emitter circuit:

R_b = dynamic base resistance
R_c = dynamic collector resistance
R_e = dynamic emitter resistance

R_{in} = input resistance
R_{rt} = reverse transfer resistance
R_{ft} = forward transfer resistance
R_{out} = output resistance

Then the following equations hold:

$$R_{in} = R_e + R_b$$
$$R_{rt} = R_e$$
$$R_{ft} = R_e - \alpha \times R_c$$
$$R_{out} = R_c + R_e - \alpha \times R_c$$

Resistance parameters (common collector)

Let α represent the dynamic current amplification of a bipolar transistor in the common-base arrangement. Let the following symbols represent resistances (in ohms) in a common-collector circuit (Figure 13.4):

R_b = dynamic base resistance
R_c = dynamic collector resistance
R_e = dynamic emitter resistance
R_{in} = input resistance
R_{rt} = reverse transfer resistance
R_{ft} = forward transfer resistance
R_{out} = output resistance

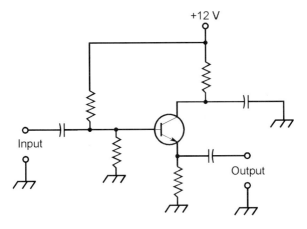

Figure 13.4 Common-collector bipolar transistor circuit.

Then the following equations hold:

$$R_{in} = R_b + R_c$$
$$R_{rt} = R_c - \alpha \times R_c$$
$$R_{ft} = R_e \times (1 - \alpha)$$
$$R_{out} = R_e + R_c - \alpha \times R_c$$

Hybrid parameters (common emitter)

In a common-emitter bipolar-transistor circuit, let the following symbols represent the indicated parameters. Currents are in amperes, resistances are in ohms, conductances are in siemens, and voltages are in volts.

I_b = base current
I_c = collector current
V_{eb} = emitter-base voltage
V_{ce} = collector-emitter voltage
R_{in} = input resistance for constant V_{ce}
G_{out} = output conductance for constant I_b
h_f = forward transfer characteristic for constant V_{ce}
h_r = reverse transfer characteristic for constant I_b

Then the following equations hold:

$$R_{in} = dV_{eb}/dI_b$$
$$G_{out} = dI_c/dV_{ce}$$
$$h_f = dI_c/dI_b$$
$$h_r = dV_{eb}/dV_{ce}$$

Hybrid parameters (common base)

In a common-base bipolar-transistor circuit, let the following symbols represent the indicated parameters. Currents are in amperes, resistances are in ohms, conductances are in siemens, and voltages are in volts.

I_e = emitter current
I_c = collector current
V_{cb} = collector-base voltage
V_{eb} = emitter-base voltage
R_{in} = input resistance for constant V_{cb}
G_{out} = output conductance for constant I_e
h_f = forward transfer characteristic for constant V_{cb}
h_r = reverse transfer characteristic for constant I_e

Then the following equations hold:

$$R_{in} = dV_{eb}/dI_e$$
$$G_{out} = dI_c/dV_{cb}$$
$$h_f = dI_c/dI_e$$
$$h_r = dV_{eb}/dV_{cb}$$

Hybrid parameters (common collector)

In a common-collector bipolar-transistor circuit, let the following symbols represent the indicated parameters. Currents are in amperes, resistances are in ohms, conductances are in siemens, and voltages are in volts.

I_b = base current
I_e = emitter current
V_{ec} = emitter-collector voltage
V_{bc} = base-collector voltage
R_{in} = input resistance for constant V_{ec}
G_{out} = output conductance for constant I_b
h_f = forward transfer characteristic for constant V_{ec}
h_r = reverse transfer characteristic for constant I_b

Then the following equations hold:

$$R_{in} = dV_{ec}/dI_e$$
$$G_{out} = dI_e/dV_{ec}$$
$$h_f = dV_{bc}/dI_b$$
$$h_r = dV_{bc}/dV_{ec}$$

Field-Effect Transistors

This section contains formulas relevant to field-effect transistors (FETs), both the N-channel type and the P-channel type.

Forward transconductance (common source)

Let I_d represent the drain current (in amperes) in a common-source FET circuit (Figure 13.5 or 13.6). Let V_g represent the gate voltage (in volts). Then the *forward transconductance*, G_{fs} (in siemens), is given by:

$$G_{fs} = dI_d / dV_g$$

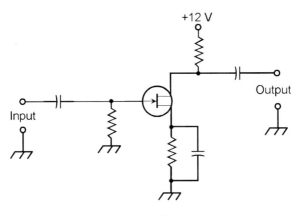

Figure 13.5 Common-source FET circuit with a bypassed source resistor.

Voltage amplification (common source)

Let G_{fs} represent the forward transconductance (in siemens) for a common-source circuit with an unbypassed source resistor (Figure 13.6). Let R_d represent the resistance (in ohms) of an external drain resistor. Let R_s represent the resistance (in ohms) of an external source resistor. Then the *voltage amplification*, A_V (as a ratio), is given by:

$$A_V = G_{fs} \times R_d / (1 + G_{fs} \times R_s)$$

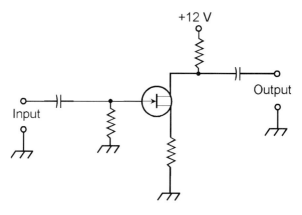

Figure 13.6 Common-source FET circuit with an unbypassed source resistor.

If the source resistor is bypassed (Figure 13.5), then:

$$A_V = G_{fs} \times R_d$$

Voltage amplification (source follower)

Let G_{fs} represent the forward transconductance (in siemens) for a source-follower FET circuit (Figure 13.7). Let R_d represent the resistance (in ohms) of an external drain resistor. Let R_s represent the resistance (in ohms) of an external

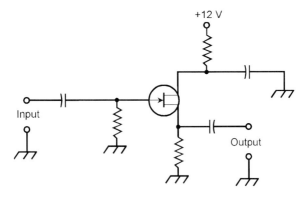

Figure 13.7 Source-follower FET circuit.

source resistor. Then the *voltage amplification*, A_V (as a ratio), is given by:

$$A_V = G_{fs} \times R_s / (1 + G_{fs} \times R_s)$$

Output impedance (source follower)

Let G_{fs} represent the forward transconductance (in siemens) for a source-follower FET circuit (Figure 13.7). Let R_s represent the resistance (in ohms) of an external source resistor. Then the *output impedance*, Z_{out} (in ohms), is given by:

$$Z_{out} = R_s / (1 + G_{fs} \times R_s)$$

Chapter 14

Electron Tubes

This chapter contains formulas relating to *electron tubes* (often simply called *tubes,* or *valves* in England). These devices have been replaced by semiconductor components in low-power applications, but tubes are still used in some applications, especially in high-power RF and audio amplifiers.

Basic Behavior

The following formulas concern fundamental relationships among the currents and voltages on the electrodes of tubes.

Diode perveance

Let A_p represent the surface area of the anode (plate) of a diode tube (in square centimeters); let s_{cp} represent the separation between the cathode and the plate (in centimeters). Then the *diode perveance*, G_d, is given by:

$$G_d = 2.3 \times 10^{-6} \times A_p/s_{cp}$$

Triode perveance

Let A_p represent the surface area of the anode (plate) of a diode tube (in square centimeters); let s_{cg} represent the separation between the cathode and the grid (in centimeters). Then the *triode perveance*, G_t, is given by:

$$G_t = 2.3 \times 10^{-6} \times A/s_{cg}$$

3/2 Power law for diode

Let V_p represent the plate voltage (in volts) in a diode tube; let G_d represent the diode perveance.

Then the plate current, I_p (in amperes) is approximately given by:

$$I_p = V_p^{3/2} \times G_d$$

3/2 Power law for triode

Let μ represent the amplification factor of a triode tube. Let V_g represent the grid voltage (in volts); let V_p represent the plate voltage (in volts); and let G_t represent the triode perveance. Then the plate current, I_p (in amperes), is approximately given by:

$$I_p = (\mu \times V_g + V_p)^{3/2} \times G_t$$

Plate current vs perveance in triode

Let V_g represent the grid voltage (in volts) in a triode; let V_p represent the plate voltage (in volts); let G_t represent the triode perveance; and let μ represent the amplification factor. Then the plate current, I_p (in amperes), is approximately given by:

$$I_p = [(\mu \times V_g + V_p)/(\mu + 1)]^{3/2} \times G_t$$

Parameters

The following formulas concern electrode resistances, electrode conductances, and amplification factors for multielement tubes.

DC internal plate resistance

Let V_p represent the DC plate-cathode voltage (in volts) in a vacuum tube. Let I_p represent the DC flowing in the plate circuit (in amperes). Then the *DC internal plate resistance*, R_p (in ohms), is given by:

$$R_p = V_p/I_p$$

Dynamic internal plate resistance

Let V_p represent the instantaneous plate-cathode voltage (in volts); let I_p represent the instantaneous current flowing in the plate circuit (in amperes). Assume that the control-grid voltage, V_g, is constant. Then the *dynamic internal plate resistance*, R_{pd} (in ohms), is given by:

$$R_{pd} = dV_p/dI_p$$

DC internal screen resistance

Let V_s represent the DC screen-grid voltage (in volts) in a tetrode or pentode tube. Let I_s represent the DC flowing in the screen circuit (in amperes). Then the *DC screen resistance*, R_s (in ohms), is given by:

$$R_s = V_s/I_s$$

Dynamic internal screen resistance

Let V_s represent the instantaneous screen-grid voltage (in volts) in a tetrode or pentode tube. Let I_s represent the instantaneous current flowing in the screen circuit (in amperes). Then the *dynamic internal screen resistance*, R_{sd} (in ohms), is given by:

$$R_{sd} = dV_s/dI_s$$

Transconductance

Let V_g represent the instantaneous control-grid voltage (in volts); let I_p represent the instantaneous current flowing in the plate circuit. Assume that the DC plate voltage, V_p, is constant. Then the *transconductance*, g_m (in siemens), is given by:

$$g_m = dI_p/dV_g$$

Plate amplification factor

Let V_p represent the instantaneous plate voltage (in volts); let V_g represent the instantaneous control-grid voltage (in volts); let g_m represent the transconductance (in siemens); let R_{pd} represent the dynamic internal plate resistance (in ohms).

Assume that the plate current, I_p, is constant. Then the *plate amplification factor*, μ_p (as a ratio), is given by either of the following formulas:

$$\mu_p = dV_p/dV_g$$
$$\mu_p = R_{pd} \times g_m$$

Screen amplification factor

Let V_s represent the instantaneous screen-grid voltage (in volts) in a tetrode or pentode tube. Let V_g represent the instantaneous control-grid voltage (in volts). Assume that the screen-grid current, I_s, is constant. Then the *screen amplification factor*, μ_s (as a ratio), is given by:

$$\mu_s = dV_s/dV_g$$

Output resistance in cathode follower

Let g_m represent the transconductance of a tube (in siemens) connected in a cathode-follower arrangement, as shown in Figure 14.1. Let R_k represent the value of the external cathode resistor (in ohms). Then the *output resistance*, R_{out} (in ohms) is given by:

$$R_{out} = R_k/(1 + g_m \times R_k)$$

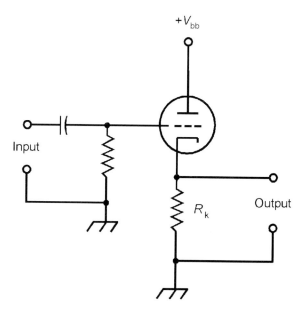

Figure 14.1 Cathode-follower electron-tube circuit.

Input capacitance

Let C_{gk} represent the capacitance between the control grid and the cathode (in picofarads); let C_{gp} represent the capacitance between the control grid and the plate (in picofarads); let µ represent the amplification factor (as a ratio). Then the *input capacitance*, C_{in} (in picofarads), is given by:

$$C_{in} = C_{gk} + C_{gp} \times (\mu + 1)$$

Circuit Formulas

The following formulas involve voltages, currents, and resistances in circuits that use electron tubes. Refer to Figure 14.2.

Required DC supply voltage

Let V_k represent the required cathode voltage (in volts); let R_k represent the value of the external cathode resistor (in ohms); let R_L represent the value of the external plate resistor (in ohms); let V_L represent the voltage (in volts) across the external plate resistor; let I_k represent the cathode current (in amperes); let I_p represent the plate current (in amperes); let V_p represent the required plate-cathode voltage (in volts). Then the *required DC supply voltage, V_{bb}* (in volts), is given by either of the following two formulas:

$$V_{bb} = V_p + V_k + V_L$$
$$V_{bb} = V_p + I_k \times R_k + I_p \times R_L$$

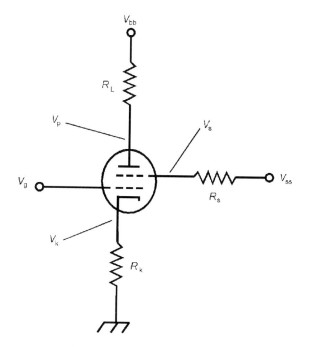

Figure 14.2 General electron-tube circuit.

DC plate-cathode voltage

Let V_{bb} represent the supply voltage (in volts); let V_k represent the voltage (in volts) on the cathode relative to ground; let I_p represent the plate current (in amperes); let R_L represent the value of the external plate resistor (in ohms). Then the *DC plate-cathode voltage*, V_p, is given by:

$$V_p = V_{bb} - (I_p \times R_L - V_k)$$

DC screen voltage

Let V_{ss} represent the DC screen-grid supply voltage (in volts); let I_s represent the screen current (in amperes); let R_s represent the value of the external screen-circuit resistor (in ohms). Then the *DC screen voltage*, V_s (in volts), is given by:

$$V_s = V_{ss} - I_s \times R_s$$

Screen current

Let V_g represent the DC voltage on the control grid (in volts); Let V_s represent the DC screen voltage (in volts); let G represent the perveance of the electron tube; let μ_s represent the screen amplification factor. Then the *screen current*, I_s (in amperes), is given by:

$$I_s = G \times (V_g + V_s/\mu_s)$$

Required external plate resistance

Let V_{bb} represent the DC supply voltage (in volts); let V_k represent the required cathode voltage (in volts) relative to ground; let V_p represent the required plate-cathode voltage (in volts); let I_p represent the required plate current (in amperes). Then the *required external plate resistance*, R_L (in ohms), is given by:

$$R_L = [V_{bb} - (V_p + V_k)]/I_p$$

Required external cathode resistance

Let V_{bb} represent the supply voltage (in volts); let V_g represent the grid voltage (in volts); let I_k represent the cathode current (in amperes). Then the *required external cathode resistance*, R_k (in ohms), is given by:

$$R_k = (V_{bb} - V_g)/I_k$$

Required external screen resistance

Let V_s represent the required screen voltage (in volts); let V_{ss} represent the screen supply voltage (in volts); let I_s represent the required screen current (in amperes). Then the *required external screen resistance*, R_s (in ohms), is given by:

$$R_s = (V_{ss} - V_s)/I_s$$

Voltage amplification and gain

Let g_m represent the transconductance of an electron tube; let μ represent the amplification factor. Let R_p represent the internal plate resistance (in ohms); let R_L represent the value of the external plate resistor (in ohms). Then the *voltage amplification*, A_V (as a ratio), is given by either of these formulas:

$$A_V = g_m \times R_p \times R_L/(R_p + R_L)$$
$$A_V = \mu \times R_L/(R_p + R_L)$$

The *voltage gain*, G_V (in decibels), is given by either of the following formulas, assuming constant impedance:

$$G_V = 20 \times \log_{10} [g_m \times R_p \times R_L/(R_p + R_L)]$$
$$G_V = 20 \times \log_{10} [\mu \times R_L/(R_p + R_L)]$$

Power Formulas

The following formulas involve electron-tube power consumption, amplification, and gain in the grounded-cathode configuration (Figure 14.2).

Power amplification and gain

Let P_{in} represent the input signal power (in watts) applied to the control grid of an electron tube; let P_{out} represent the output signal power in the plate circuit. Then the *power amplification*, A_P (as a ratio), is given by:

$$A_P = P_{out}/P_{in}$$

The *power gain*, G_P (in decibels), is given by:

$$G_P = 10 \times \log_{10}(P_{out}/P_{in})$$

Filament power demand

Let V_f represent the effective filament voltage (in volts rms); let I_f represent the effective filament current (in amperes rms); let R_f represent the filament resistance (in ohms). Then the *filament power demand*, P_f (in watts), is given by any of the following formulas:

$$P_f = V_f \times I_f$$
$$P_f = I_f^2 \times R_f$$
$$P_f = V_f^2/R_f$$

DC screen power

Let V_s represent the DC screen voltage (in volts); let I_s represent the direct current in the screen circuit (in amperes). Then the *DC screen power*, P_s (in watts), is given by:

$$P_s = V_s \times I_s$$

DC plate input power

Let V_p represent the DC plate voltage (in volts); let I_p represent the direct current in the plate circuit (in amperes). Then the *DC plate input power*, $P_{p\text{-in}}$ (in watts), is given by:

$$P_{p-in} = V_p \times I_p$$

Signal output power

Let V_{max} represent the maximum instantaneous plate voltage (in volts); let V_{min} represent the minimum instantaneous plate voltage (in volts); let I_{max} represent the maximum instantaneous plate current (in amperes); let I_{min} represent the minimum instantaneous plate current (in amperes). Then the *signal output power*, $P_{s\text{-out}}$ (in watts), is given by:

$$P_{s\text{-out}} = 0.125 \times (V_{max} \times I_{max} - V_{max} \times I_{min} - V_{min} \times I_{max} + V_{min} \times I_{min})$$

Plate power dissipation

Let $P_{\text{p-in}}$ represent the DC plate power input (in watts); let $P_{\text{s-out}}$ represent the signal output power (in watts). Then the *plate power dissipation*, $P_{p\text{-}dis}$ (in watts), is given by:

$$P_{\text{p-dis}} = P_{\text{p-in}} - P_{\text{s-out}}$$

Plate efficiency

Let $P_{\text{p-in}}$ represent the DC plate input power (in watts); let $P_{\text{s-out}}$ represent the signal output power (in watts). Then the *plate efficiency*, η_{p} (as a ratio), is given by:

$$\eta_{\text{p}} = P_{\text{s-out}}/P_{\text{p-in}}$$

As a percentage, the plate efficiency $\eta_{\text{p\%}}$ is given by:

$$\eta_{\text{p\%}} = 100 \times P_{\text{s-out}}/P_{\text{p-in}}$$

Input power sensitivity

Let $V_{\text{g-in}}$ represent the grid input signal voltage (in volts rms); let $P_{\text{s-out}}$ represent the signal power output (in watts). Then the *input power sensitivity*, S_{P} (in watts per volt), is given by:

$$S_{\text{P}} = P_{\text{s-out}}/V_{\text{g-in}}$$

Chapter 15

Electromagnetic Waves and Antenna Systems

This chapter contains formulas and data involving electromagnetic fields, transmission lines, and antennas.

Electromagnetic Fields

An *electromagnetic (EM) field* is generated whenever charged particles are accelerated. In most practical situations, this acceleration is alternating and periodic.

Frequency vs. wavelength

Let the frequency (in hertz) of an EM wave be represented by f; let the wavelength (in meters) be represented by λ; let the speed of propagation (in meters per second) be represented by c. Then the following formula holds:

$$c = f \times \lambda$$

In free space, c is approximately 2.99792×10^8 meters per second. For most practical applications this is rounded off to 3.00×10^8 meters per second.

Free-space wavelength

The *free-space wavelength* of an RF field depends on the frequency. In general, the higher the frequency, the shorter the free-space wavelength. Let:

λ_{ft} = free-space wavelength (in feet)
λ_{in} = free-space wavelength (in inches)
λ_{m} = free-space wavelength (in meters)
λ_{cm} = free-space wavelength (in centimeters)
f_{MHz} = frequency (in megahertz)
f_{GHz} = frequency (in gigahertz)

Electromagnetic Waves and Antenna Systems

Then the following equations hold:

$$\lambda_{ft} = 984/f_{MHz}$$
$$\lambda_{ft} = 0.984/f_{GHz}$$
$$\lambda_{in} = 11.8/f_{GHz}$$
$$\lambda_{m} = 300/f_{MHz}$$
$$\lambda_{m} = 0.300/f_{GHz}$$
$$\lambda_{cm} = 30.0/f_{GHz}$$

Angular frequency

Let f be the frequency of an EM field (in hertz). Then the *angular frequency*, ω (in radians per second), is given by:

$$\omega = 2 \times \pi \times f \approx 6.28 \times f$$

The angular frequency in degrees per second is given by:

$$\omega = 360 \times f$$

Period

Let f be the frequency of an EM field (in hertz). The the period, T (in seconds), is given by:

$$T = 1/f$$

For an angular frequency ω in radians per second:

$$T = 2 \times \pi/\omega \approx 6.28/\omega$$

For an angular frequency ω in degrees per second:

$$T = 360/\omega$$

RF Transmission Lines

The most common types of RF transmission line are *coaxial cable* (unbalanced) and *two-wire line* (balanced). The following formulas apply to such lines with a dielectric consisting of dry air.

Characteristic impedance of coaxial cable

Let d_1 represent the outside diameter of the center conductor in a coaxial transmission line; let d_2 represent the inside diameter of the shield or braid (in the same units as d_1). Then the *characteristic impedance*, Z_0, of the line (in ohms), is given by:

$$Z_0 = 138 \times \log_{10}(d_2/d_1)$$

Characteristic impedance of two-wire line

Let d represent the outside diameter of either wire in a two-wire transmission line; assume both

Electromagnetic Waves and Antenna Systems

wires have the same diameter. Let s represent the spacing between the centers of the two conductors in the line; assume that s is the same at all points along the line and is specified in the same units as d. Then the characteristic impedance, Z_0, of the line (in ohms), is given by:

$$Z_0 = 276 \times \log_{10}(2 \times s/d)$$

Velocity factor

Let c_0 represent the speed at which an EM disturbance propagates along a transmission line (in meters per second). Then the *velocity factor, v,* of the line (as a ratio), is given by:

$$v = c_0/(3.00 \times 10^8)$$

The velocity factor as a percentage is denoted $v_\%$ and is given by:

$$v_\% = c_0/(3.00 \times 10^6)$$

Table 15.1 lists approximate velocity factors for common types of RF transmission line.

Electrical wavelength

In a medium other than free space, the wavelength depends on the frequency and also on the

TABLE 15.1 Velocity Factors for RF Transmission Lines. These Figures are Approximate

General Description	Velocity Factor
Coaxial cable, solid polyethylene dielectric	0.66
Coaxial hard line, solid polyethylene dielectric	0.66
Coaxial cable, foamed polyethylene dielectric	0.75 – 0.85
Coaxial hard line, foamed polyethylene dielectric	0.75 – 0.85
Coaxial hard line, solid polyethylene disk spacers	0.85 – 0.90
TV "twin-lead" ribbon, 75-ohm	0.70 – 0.80
TV "twin-lead" ribbon, 300-ohm	0.80 – 0.90
Parallel-wire "window" ribbon	0.85 – 0.90
Parallel-wire "ladder line" with plastic spacers	0.90 – 0.95
Open-wire lines without spacers	0.95
Single-wire line	0.95

velocity factor (v) of the medium in which the field propagates. Let:

λ_{ft} = electrical wavelength (in feet)

λ_{in} = electrical wavelength (in inches)

λ_{m} = electrical wavelength (in meters)

λ_{cm} = electrical wavelength (in centimeters)

f_{MHz} = frequency (in megahertz)

f_{GHz} = frequency (in gigahertz)

Electromagnetic Waves and Antenna Systems

Then the following equations hold:

$$\lambda_{ft} = 984 \times v/f_{MHz}$$
$$\lambda_{ft} = 0.984 \times v/f_{GHz}$$
$$\lambda_{in} = 11.8 \times v/f_{GHz}$$
$$\lambda_{m} = 300 \times v/f_{MHz}$$
$$\lambda_{m} = 0.300 \times v/f_{GHz}$$
$$\lambda_{cm} = 30.0 \times v/f_{GHz}$$

Let $v_\%$ represent the velocity factor as a percentage between 0 and 100. Then:

$$\lambda_{ft} = 9.84 \times v_\%/f_{MHz}$$
$$\lambda_{ft} = 9.84 \times 10^{-3} \times v_\%/f_{GHz}$$
$$\lambda_{in} = 0.118 \times v_\%/f_{GHz}$$
$$\lambda_{m} = 3.00 \times v_\%/f_{MHz}$$
$$\lambda_{m} = 3.00 \times 10^{-3} \times v_\%/f_{GHz}$$
$$\lambda_{cm} = 0.300 \times v_\%/f_{GHz}$$

Length of 1/4-wave matching section

The length of a quarter-wave section of transmission line, commonly used for impedance matching, depends on the frequency, and also on the velocity factor of the line. Let:

$$s_{ft} = \text{section length (in feet)}$$
$$s_{in} = \text{section length (in inches)}$$

Chapter Fifteen

s_m = section length (in meters)
s_{cm} = section length (in centimeters)
f_{MHz} = frequency (in megahertz)
f_{GHz} = frequency (in gigahertz)
v = velocity factor (as a ratio between 0 and 1)

Then the following equations hold:

$$s_{cm} = 7.50 \times v/f_{GHz}$$
$$s_{ft} = 246 \times v/f_{MHz}$$
$$s_{ft} = 0.246 \times v/f_{GHz}$$
$$s_{in} = 2.95 \times v/f_{GHz}$$
$$s_m = 75.0 \times v/f_{MHz}$$
$$s_m = 7.50 \times 10^{-2} \times v/f_{GHz}$$

Let $v_\%$ represent the velocity factor as a percentage between 0 and 100. Then:

$$s_{cm} = 7.50 \times 10^{-2} \, v_\%/f_{GHz}$$
$$s_{ft} = 2.46 \times v_\%/f_{MHz}$$
$$s_{ft} = 2.46 \times 10^{-3} \times v_\%/f_{GHz}$$
$$s_{in} = 2.95 \times 10^{-2} \times v_\%/f_{GHz}$$
$$s_m = 0.750 \times v_\%/f_{MHz}$$
$$s_m = 7.50 \times 10^{-4} \times v_\%/f_{GHz}$$

Characteristic impedance of 1/4-wave matching section

The *characteristic impedance* of a quarter-wave matching section must be equal to the geometric mean of the input and output impedances. Let:

Z_0 = characteristic impedance of matching section (in ohms)

Z_{in} = input impedance (in ohms)

Z_{out} = output impedance (in ohms)

The following formula applies:

$$Z_0 = (Z_{in} \times Z_{out})^{1/2}$$

Standing wave ratio (SWR)

Suppose that an RF transmission line is terminated in a load whose impedance (in ohms) is a pure resistance, R_{load}. Let Z_0 represent the characteristic impedance of the line (in ohms). If $R_{load} > Z_0$, the *standing-wave ratio,* abbreviated SWR, is given by:

$$SWR = R_{load}/Z_0$$

If $R_{load} < Z_0$, then:

$$SWR = Z_0/R_{load}$$

If $R_{\text{load}} = Z_0$, then:

$$\text{SWR} = R_{\text{load}}/Z_0 = Z_0/R_{\text{load}} = 1:1$$

When a transmission line is terminated with a load whose impedance is not a pure resistance, the SWR is determined according to the maximum and minimum voltage or current in the line.

Voltage standing wave ratio (VSWR)

Let $V_{\max}$ represent the maximum RF voltage (in volts) between the conductors of a transmission line; let $V_{\min}$ represent the minimum RF voltage (in volts) between the conductors of the line. Points at which $V_{\max}$ and $V_{\min}$ occur are separated by 1/4 electrical wavelength. The *voltage standing-wave ratio*, abbreviated *VSWR*, is given by:

$$\text{VSWR} = V_{\max}/V_{\min}$$

Current standing wave ratio (ISWR)

Let $I_{\max}$ represent the maximum RF current (in amperes) in a transmission line; let $I_{\min}$ represent the minimum RF current in the line (in amperes). Points at which $I_{\max}$ and $I_{\min}$ occur are separated by a 1/4 electrical wavelength. Current maxima normally exist at the same points on a transmission line as voltage minima; current minima normally exist at the same points as voltage maxima. The *current standing-wave ratio,* abbreviated ISWR, is given by:

$$\text{ISWR} = I_{\max}/I_{\min}$$

Relationship among SWR, VSWR, and ISWR

In theory, assuming zero loss in a transmission line, the following equation holds:

$$\text{SWR} = \text{VSWR} = \text{ISWR}$$

In practice, when a line has significant loss, these quantities differ slightly, depending on the points where current and voltage are measured. The ratios are lower toward the equipment (transmitter) end of the line, and higher toward the antenna (load) end.

Reflection coefficient vs SWR

Let s represent the SWR, VSWR, or ISWR measured at the antenna (load) end of an RF transmission line. Then the *reflection coefficient*, k, is given by:

$$k = (s - 1)/s$$

Reflection coefficient vs load resistance

Suppose that an RF transmission line is terminated in a load whose impedance (in ohms) is a pure resistance, R_{load}. Let Z_0 represent the characteristic impedance of the line (in ohms). Then the *reflection coefficient, k,* is given by:

$$k = (R_{load} - Z_0)/(R_{load} + Z_0)$$

Loss in matched lines

Table 15.2 gives the approximate loss (in decibels per 100 feet and per 100 meters) for various types

TABLE 15.2A Approximate Loss in Decibels Per 100 Feet for Various Transmission Lines Under Conditions of 1:1 SWR.

Line type	1 MHz	10 MHz	100 MHz
600-ohm ladder line	0.05	0.1	0.5
300-ohm TV ribbon	0.1	0.5	1.5
RG-8/U coaxial cable	0.15	0.6	2.0
RG-59/U coaxial cable	0.3	1.0	4.0
RG-58/U coaxial cable	0.3	1.4	5.0

TABLE 15.2B Approximate Loss in Decibels Per 100 Meters for Various Transmission Lines Under Conditions of 1:1 SWR.

Line type	1 MHz	10 MHz	100 MHz
600-ohm ladder line	0.16	0.33	1.6
300-ohm TV ribbon	0.33	3.3	4.9
RG-8/U coaxial cable	0.49	2.0	6.5
RG-59/U coaxial cable	1.0	3.3	1.3
RG-58/U coaxial cable	0.3	1.4	5.0

of transmission line under conditions of 1:1 SWR (a perfect match). Dielectrics are assumed to be solid polyethylene, except for ladder line, in which the dielectric is dry air with plastic spacers.

SWR loss

Figure 15.1 shows the approximate loss (in decibels) that occurs in addition to the matched-line loss in a transmission line when the SWR is not 1:1. This additional loss is called *SWR loss* and is minimal unless the SWR is more than 2:1. In severely mismatched, long lines at high frequencies, the SWR loss can be considerable.

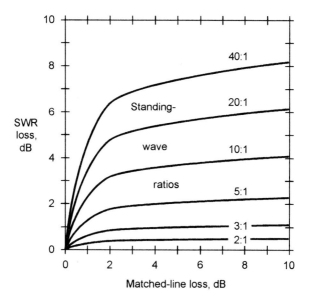

Figure 15.1 Approximate SWR loss as a function of matched-line loss and SWR (as measured at the load end of the line).

Antennas

The physical size of a resonant antenna depends on the electrical wavelength, which, in turn, depends on the frequency.

Electromagnetic Waves and Antenna Systems

Radiation resistance

Let P_{rad} represent the power radiated from a resonant antenna (in watts); let I_{rad} represent the current (in amperes) that would flow in a resistor inserted at the feed point, if the use of that resistor in place of the antenna would produce the same feed-line current distribution as the antenna. Then the *radiation resistance,* R_{rad} (in ohms), of the resonant antenna is given by:

$$R_{rad} = P_{rad}/I_{rad}^2$$

Antenna efficiency

Let R_{rad} represent the radiation resistance of an antenna (in ohms); let R_{loss} represent the *loss resistance* in the antenna and associated components, such as loading coils, traps, ground system, etc. Then the *antenna efficiency,* η (as a ratio), is given by:

$$\eta = R_{rad}/(R_{rad} + R_{loss})$$

The efficiency as a percentage, $\eta_\%$, is given by the following formula:

$$\eta_\% = 100 \times R_{rad}/(R_{rad} + R_{loss})$$

Length of 1/2-wave dipole antenna

For a half-wave dipole antenna fed at the center, placed at least 1/4 wavelength above effective ground and constructed of common wire, let:

s_{ft} = end-to-end length (in feet)
s_{in} = end-to-end length (in inches)
s_m = end-to-end length (in meters)
s_{cm} = end-to-end length (in centimeters)
f_{MHz} = frequency (in megahertz)
f_{GHz} = frequency (in gigahertz)

Then the following formulas apply:

$$s_{ft} = 468/f_{MHz}$$
$$s_{ft} = 0.468/f_{GHz}$$
$$s_{in} = 5.62/f_{GHz}$$
$$s_m = 143/f_{MHz}$$
$$s_m = 0.143/f_{GHz}$$
$$s_{cm} = 14.3/f_{GHz}$$

For antennas constructed of metal tubing, these values should be multiplied by approximately 0.95 (95 percent). However, the exact optimum antenna length in any given case must be determined by experimentation because it depends on

Electromagnetic Waves and Antenna Systems

the ratio of tubing diameter to wavelength and also on the surrounding environment.

Height of 1/4-wave vertical antenna

For a quarter-wave vertical antenna constructed of common wire and placed over perfectly conducting ground, let:

h_{ft} = radiating-element height (in feet)
h_{in} = radiating-element height (in inches)
h_m = radiating-element height (in meters)
h_{cm} = radiating-element height (in centimeters)
f_{MHz} = frequency (in megahertz)
f_{GHz} = frequency (in gigahertz)

Then the following formulas apply:

$$h_{ft} = 234/f_{MHz}$$
$$h_{ft} = 0.234/f_{GHz}$$
$$h_{in} = 2.81/f_{GHz}$$
$$h_m = 71.5/f_{MHz}$$
$$h_m = 7.15 \times 10^{-2}/f_{GHz}$$
$$h_{cm} = 7.15/f_{GHz}$$

For antennas constructed of metal tubing, the above values should be multiplied by approximately 0.95 (95 percent). However, the exact optimum antenna height in any given case must be determined by experimentation because it depends on the ratio of tubing diameter to wavelength and also on the surrounding environment.

Length of resonant harmonic antenna

For a resonant harmonic antenna fed at integral multiples of 1/4 wavelength from either end, placed at least 1/4 wavelength above effective ground and constructed of common wire, let:

s_{ft} = end-to-end length (in feet)

s_{in} = end-to-end length (in inches)

s_m = end-to-end length (in meters)

s_{cm} = end-to-end length (in centimeters)

f_{MHz} = frequency (in megahertz)

f_{GHz} = frequency (in gigahertz)

n = harmonic at which antenna is operated (a positive integer)

Then the following formulas apply:

$$s_{ft} = 492 \times (n - 0.05)/f_{MHz}$$
$$s_{ft} = 0.492 \times (n - 0.05)/f_{GHz}$$
$$s_{in} = 5.90 \times (n - 0.05)/f_{GHz}$$
$$s_m = 150 \times (n - 0.05)/f_{MHz}$$

Electromagnetic Waves and Antenna Systems

$$s_\mathrm{m} = 0.150 \times (n - 0.05)/f_\mathrm{GHz}$$
$$s_\mathrm{cm} = 15.0 \times (n - 0.05)/f_\mathrm{GHz}$$

Length of resonant unterminated long wire

For a resonant unterminated long wire antenna fed at either end, placed at least 1/4 wavelength above effective ground, and constructed of common wire, let:

s_ft = end-to-end length (in feet)

s_in = end-to-end length (in inches)

s_m = end-to-end length (in meters)

s_cm = end-to-end length (in centimeters)

f_MHz = frequency (in megahertz)

f_GHz = frequency (in gigahertz)

n = length of wire in wavelengths

Then the following formulas apply:

$$s_\mathrm{ft} = 984 \times (n - 0.025)/f_\mathrm{MHz}$$
$$s_\mathrm{ft} = 0.984 \times (n - 0.025)/f_\mathrm{GHz}$$
$$s_\mathrm{in} = 11.8 \times (n - 0.025)/f_\mathrm{GHz}$$
$$s_\mathrm{m} = 300 \times (n - 0.025)/f_\mathrm{MHz}$$
$$s_\mathrm{m} = 0.300 \times (n - 0.025)/f_\mathrm{GHz}$$
$$s_\mathrm{cm} = 30.0 \times (n - 0.025)/f_\mathrm{GHz}$$

Chapter 16

Measurement

This chapter contains formulas and diagrams relevant to bridge circuits, null networks, error estimation, and the interpolation of measured data.

Bridge Circuits

A *bridge circuit* is used to determine unknown resistances, reactances, impedances, and/or frequencies. Variable components are adjusted until a condition of balance (zero output) occurs, at which time the unknown values can be calculated.

Anderson bridge

Let L_x and R_x represent an unknown inductance (in henrys) and an unknown resistance (in ohms) in series. Assume that they are inserted in the *Anderson bridge* configuration of Figure 16.1 and that the variable components are adjusted for balance. Let C_s represent a precision standard capacitance (in farads). The following formulas apply:

$$L_x = C_s \times [R_3 \times (1 + R_2/R_4) + R_2]$$
$$R_x = R_1 \times R_2/R_4$$

Hay bridge

Let L_x and R_x represent an unknown inductance (in henrys) and an unknown resistance (in ohms) in series. Assume that they are inserted in the *Hay bridge* configuration of Figure 16.2, and that the variable components are adjusted for balance. Let C_s represent a precision standard capacitance (in farads). Let f represent the frequency (in hertz). The following formulas apply:

$$L_x = C_s \times R_1 \times R_2$$
$$R_x = (4 \times \pi^2 \times f^2 \times C_s^2 \times R_1 \times R_2 \times R_3)/$$
$$(1 + 4 \times \pi^2 \times f^2 \times C_s^2 \times R_3^2)$$

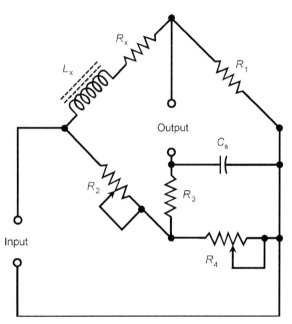

Figure 16.1 Anderson bridge for determining the value of an inductance (L_x) and resistance (R_x) in series.

Maxwell bridge

Let L_x and R_x represent an unknown inductance (in henrys) and an unknown resistance (in ohms) in series. Assume that they are inserted in the *Maxwell bridge* configuration of Figure 16.3, and

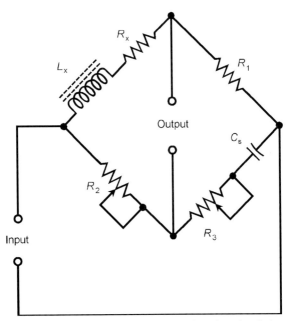

Figure 16.2 Hay bridge for determining the value of an inductance (L_x) and resistance (R_x) in series.

that the variable components are adjusted for balance. Let C_s represent a precision standard capacitance (in farads). The following formulas apply:

$$L_x = C_s \times R_1 \times R_2$$
$$R_x = R_1 \times R_2/R_3$$

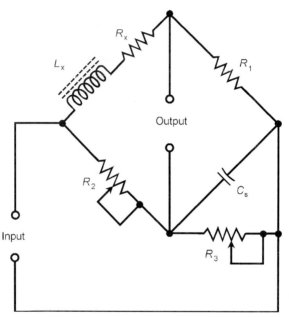

Figure 16.3 Maxwell bridge for determining the value of an inductance (L_x) and resistance (R_x) in series.

Owen bridge

Let L_x and R_x represent an unknown inductance (in henrys) and an unknown resistance (in ohms) in series. Assume that they are inserted in the *Owen bridge* configuration of Figure 16.4 and

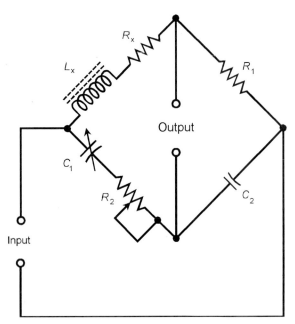

Figure 16.4 Owen bridge for determining the value of an inductance (L_x) and resistance (R_x) in series.

that the variable components are adjusted for balance. The following formulas apply:

$$L_x = C_2 \times R_1 \times R_2$$
$$R_x = R_1 \times C_2/C_1$$

Schering bridge

Let C_x and R_x represent an unknown capacitance (in farads) and an unknown resistance (in ohms) in series. Assume that they are inserted in the *Schering bridge* configuration of Figure 16.5, and

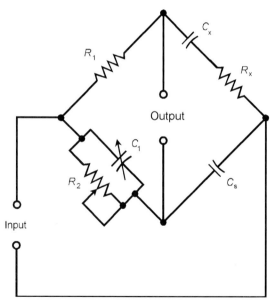

Figure 16.5 Schering bridge for determining the value of a capacitance (C_x) and resistance (R_x) in series.

that the variable components are adjusted for balance. Let C_s represent a precision standard capacitance (in farads). The following formulas apply:

$$C_x = C_s \times R_2/R_1$$
$$R_x = R_1 \times C_1/C_s$$

Wheatstone bridge

Let R_x represent an unknown resistance (in ohms). Assume that it is inserted in the *Wheatstone bridge* configuration of Figure 16.6, and that potentiometer R_2 is adjusted for balance. The following formula applies:

$$R_x = R_1 \times R_2/R_3$$

Wien bridge

Let the resistances (in ohms) and capacitances (in farads) of the *Wien bridge* circuit of Figure 16.7 be related as follows:

$$R_2 = 2 \times R_1$$
$$C_1 = C_2$$
$$R_3 = R_4$$

Then the input frequency, f (in hertz), that results in zero output (balance) is given by:

$$f = 1/(2 \times \pi \times R_3 \times C_1)$$

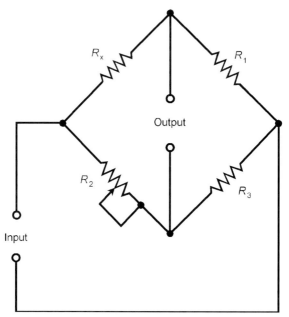

Figure 16.6 Wheatstone bridge for determining the value of a resistance (R_x).

Null Networks

A *null network* produces zero output at a specific frequency that is determined by the values of the inductances, capacitances, and resistances in the circuit.

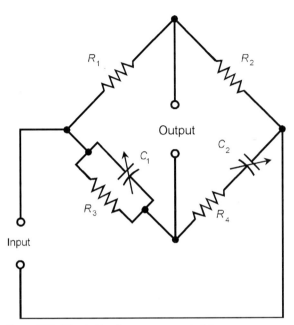

Figure 16.7 Wien bridge for measurement of frequency.

LC bridged *T*

Suppose that an inductance, L (in henrys); capacitances, C_1 and C_2 (in farads); and a resistance, R (in ohms); are connected in the configuration shown in Figure 16.8. Further, suppose that $C_1 = C_2$, so the network is symmetrical. Then the null frequency, f (in hertz), is given by either of the following:

$$f = 1/[\pi \times (2 \times L \times C_1)^{1/2}]$$
$$f = 1/[\pi \times (2 \times L \times C_2)^{1/2}]$$

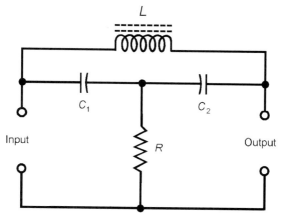

Figure 16.8 Inductance-capacitance (LC) bridged T null network.

RC bridged T

Suppose that capacitances, C_1 and C_2 (in farads), and resistances, R_1 and R_2 (in ohms), are connected in the configuration, as shown in Figure 16.9. Further suppose that $C_1 = C_2$, so the network is symmetrical. Then the null frequency, f (in hertz), is given by either of the following:

$$f = 1/[2 \times \pi \times C_1 \times (R_1 \times R_2)^{1/2}]$$
$$f = 1/[2 \times \pi \times C_2 \times (R_1 \times R_2)^{1/2}]$$

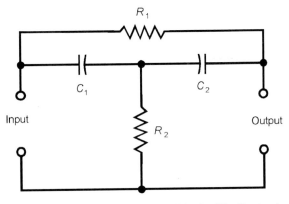

Figure 16.9 Resistance-capacitance (RC) bridged T null network.

RC parallel *T*

Suppose capacitances C_1, C_2, and C_3 (in farads) and resistances R_1, R_2, and R_3 (in ohms) are connected in the configuration shown in Figure 16.10. Further suppose that the following relationships hold:

$$C_3 = 2 \times C_1 = 2 \times C_2$$
$$R_1 = R_2 = 2 \times R_3$$

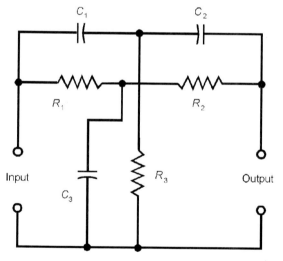

Figure 16.10 Resistance-capacitance (RC) parallel T null network.

Then the null frequency, f (in hertz), is given by:

$$f = 1/(2 \times \pi \times R_1 \times C_1)$$

Error and Interpolation

Measurement error

Let x_a represent the actual value of a quantity to be measured. Let x_m represent the measured value of the quantity, in the same units as x_a. Then the *absolute error*, D_a (in the same units as x_a), is given by:

$$D_a = x_m - x_a$$

The *proportional error*, D_p, is given by:

$$D_p = (x_m - x_a)/x_a$$

The *percentage error*, $D_\%$, is given by:

$$D_\% = 100 \times (x_m - x_a)/x_a$$

Error values and percentages are positive if $x_m > x_a$, and negative if $x_m < x_a$.

Arithmetic interpolation

Let $y = f(x)$ represent a function in which the value of a quantity (y) depends on the value of an independent variable (x). Let y_1 and y_2 represent two values of the function such that:

$$y_1 = f(x_1)$$
$$y_2 = f(x_2)$$

Then the value y_a of the function at a point x_m midway between x_1 and x_2 can be approximated via arithmetic interpolation:

$$y_a = (y_1 + y_2)/2 = [f(x_1) + f(x_2)]/2$$

Geometric interpolation

Let $y = f(x)$ represent a function in which the value of a quantity (y) depends on the value of an independent variable (x). Let y_1 and y_2 represent two values of the function such that:

$$y_1 = f(x_1)$$
$$y_2 = f(x_2)$$

Then the value y_g of the function at a point (x_m) midway between x_1 and x_2 can be approximated via *geometric interpolation:*

$$y_g = (y_1 \times y_2)^{1/2} = [f(x_1) \times f(x_2)]^{1/2}$$

Bibliography

Crowhurst, N. and Gibilisco, S., *Mastering Technical Mathematics—2nd Edition* (New York, NY: McGraw-Hill, 1999)

Dorf, R. *Electrical Engineering Handbook—2nd Edition* (Boca Raton, FL: CRC Press, 1997)

Gibilisco, S., *Electronics Portable Handbook* (New York, NY: McGraw-Hill, 1999)

Gibilisco, S., *Handbook of Radio and Wireless Technology* (New York, NY: McGraw-Hill, 1999)

Gibilisco, S., *Teach Yourself Electricity and Electronics—2nd Edition* (New York, NY: McGraw-Hill, 1997)

Gibilisco, S., *The Illustrated Dictionary of Electronics—7th Edition* (New York, NY: McGraw-Hill, 1997)

Van Valkenburg, M., *Reference Data for Engineers: Radio, Electronics, Computer and Communications* (Indianapolis, IN: Howard W. Sams & Co., 1998)

Veley, V., *The Benchtop Electronics Reference Manual* (New York, NY: McGraw-Hill, 1994).

Index

1/4-wave matching section, 273–275
1/4-wave vertical antenna, 283–284
1/2-wave dipole antenna, 282–283

abscissa, 67
absolute value
    impedance, 10
    of complex number, 159
AC
    amperage, 166–168
    amplitude expressions, 154–158
    conductance, 162–163
    energy, 172–174
    power, 170–172
    voltage, 168–170
acceleration, 18
    angular, 18
    linear, 18
admittance, 162–166
air cavity
    1/4 wave, 206
    1/2 wave, 207
algebra, 65–77
alpha, 237, 239
alternating current, 151–174
alternative unit systems, 22
amperage
    AC, 166–168
    DC, 136–140
ampere, 3, 136–137
ampere rms, 166
ampere-meter, 13
ampere-turn, 13
ampere-turn per meter, 14
ampere-turn per weber, 13, 176
amplitude
    average, 156
    effective, 157
    instantaneous, 155, 157
    negative peak, 156
    peak, 156
    peak-to-peak, 157

Index

positive peak, 156
root-mean-square (rms), 157
AND gate, 196, 198
AND operation, 194
Anderson bridge, 288, 289
angular acceleration, 18
angular frequency, 269
angular measure
    plane, 17
    solid, 17
antenna, 280–285
    1/4-wave vertical, 283–284
    1/2-wave dipole, 282–283
    efficiency, 281
    radiation resistance, 281
    resonant harmonic, 284–285
    resonant unterminated long wire, 285
antiderivative, 124
anti-proton, 4
apparent energy, 173–174
apparent power, 171–172
area, 16
arithmetic series, 89
arithmetic-geometric series, 91
arithmetic interpolation, 300–301
asynchronous flip-flop, 201
average amplitude, 156
Avogadro's number, 3

bandpass filters, 217–222
    constant-k, 218–219
    series m-derived, 219–220, 221
    shunt m-derived, 220–222
bandstop filters, 222–226
    constant-k, 222–224
    series m-derived, 224–225, 226
    shunt m-derived, 225–227
basic binary operations, 193–195
basic series, 89–91
beta, 237, 239
binary numbers, 188–189
binary operations, 193–196
    basic, 193–195
    secondary, 195–196
bipolar transistor, 235–245
    alpha, 237, 239
    beta, 238, 239
    dynamic base resistance, 236
    dynamic collector resistance, 236–237
    dynamic emitter feedback conductance, 237
    dynamic emitter resistance, 236
    dynamic stability factor, 239

hybrid parameters
(common base),
243–244
hybrid parameters
(common collector),
244–245
hybrid parameters
(common emitter),
242–243
resistance parameters
(common base),
239–240
resistance parameters
(common collector),
241–242
resistance parameters
(common emitter),
240–241
static forward
current transfer
ratio, 235
bridge circuits, 287–296
Anderson, 288, 289
Hay, 288, 290
Maxwell, 289–291
Owen, 291–292
Schering, 293–294
Wheatstone, 294, 295
Wien, 294, 296
Boolean theorems, 199–200

candela, 3
capacitance, 8
capacitive reactance, 161

capacitive susceptance, 164
carrier mobility, 11–12
Cartesian plane, 67, 77
Cartesian three-space, 73–74
cathode-follower electron-
tube circuit, 257
celestial coordinates, 72–73
celestial latitude, 72
celestial longitude, 72
Centimeter/Gram/Second
System, 22
characteristic curve, 234
characteristic impedance
coaxial cable, 270
two-wire line, 270–271
charge, DC, 135–136
charge quantity, electrical, 4
charge-carrier mobility, 11–12
charging, 137–138
coincident sets, 100
common-base bipolar
transistor circuit, 238
common-collector bipolar
transistor circuit, 242
common-emitter bipolar
transistor circuit, 235
common-source FET circuit
with bypassed source
resistor, 246
with unbypassed source
resistor, 247
common logarithms, 82
complex admittances in
parallel, 165
complex impedance, 10, 158

complex impedances in parallel, 165–166
complex impedances in series, 162
complex numbers, 158–159
conditionally convergent series, 88
conductance, 6
conductivity, 6
constant of integration, 125
constants, 35, 41–42
continuity, function, 104
continuous function, 104, 105
convergent series, 87
conversions, 21–40
    electrical unit, 28, 29–33
    magnetic unit, 28, 34–35
    miscellaneous unit, 35, 36–40
    SI unit, 24–27
coordinate systems, 67–77
cosecant, 78
cosine, 77
cotangent, 78
coulomb, 4, 135
coulomb per volt-meter, 11
Coulomb's law, 136
cross product, of vectors, 112, 113
cubic function, 106
cubic meter, 16
current, electric, 3
current gain, 15
current loss, 15
current standing-wave ratio, 276–277
cycle per second, 7
cylindrical coordinates, 74–76

DC amperage, 136–140
DC charge, 135–136
DC component, 156
DC energy, 147–149
DC internal plate resistance, 254
DC internal screen resistance, 254
DC plate-cathode voltage, 259
DC plate input power, 264
DC power, 146–147
DC resistance, 144–146
DC screen power, 263–264
DC screen voltage, 260
DC voltage, 140–144
decibel, 14–16
decimal, 188
declination, 72
definite integral, 125, 126
degree Kelvin, 2–3
derivatives, 115–121
    function raised to a power, 119
    higher-order, 117
    multiplication by a constant, 118
    product of, 118
    quotient of, 118

Index

second, 117
sum and difference of, 117–118
differentiator, 121–124
digital electronics, 187–203
diode, 231–234
    3/2 power law for, 252–253
    characteristic curve, 234
    dynamic resistance, 233
    forward current, 232
    perveance, 252
    rectification efficiency, 233
    static resistance, 232–233
direct current, 135–149
direction angles, 113
direction cosines, 113
discharging, 137–138
discontinuity, function, 104, 105
discontinuous function, 104, 105
disjoint sets, 99–100
displacement, 2
divergent series, 87
domain, function, 102, 103
dot product, of vectors, 112, 113
dynamic base resistance, 236
dynamic collector resistance, 236–237
dynamic emitter feedback conductance, 237
dynamic emitter resistance, 236
dynamic internal plate resistance, 254
dynamic internal screen resistance, 255
dynamic resistance, 233
dynamic stability factor, 239

eddy-current loss, 184
effective amplitude, 157
electric charge quantity, 4
electric current, 3
electric field strength, 10–11
electric potential, 5
electric susceptibility, 11
electrical unit conversions, 28, 29–33
electrical units, 4–12
electrical wavelength, 271–273
electromagnetic field, 267–270
electromagnetic field strength, 11
electromotive force, 5, 140
electron, 4
electron tube, 251–265
    circuit, general, 259
    DC internal plate resistance, 254

electron tube, *continued*
- DC internal screen resistance, 254
- DC plate-cathode voltage, 259
- DC plate input power, 264
- DC screen power, 263–264
- DC screen voltage, 260
- dynamic internal plate resistance, 254
- dynamic internal screen resistance, 255
- filament power demand, 263
- input capacitance, 257
- input power sensitivity, 265
- output resistance in cathode follower, 256
- plate amplification factor, 255–256
- plate efficiency, 265
- plate power dissipation, 265
- power amplification and gain, 263
- required DC supply voltage, 258
- required external cathode resistance, 261
- required external plate resistance, 261
- required external screen resistance, 261–262
- screen amplification factor, 256
- screen current, 260
- signal output power, 264
- transconductance, 255
- voltage amplification and gain, 262

element, set, 97–98
energy, 4–5
- AC, 172–174
- apparent, 173–174
- DC, 147–149
- reactive, 173
- real, 172–173

equatorial axis, 71
error, measurement, 300
exponential function, 107
exponential series, 93–94

factorial, 88
farad, 8
farad per meter, 11
field-effect transistor, 245–249
- forward transconductance (common source), 246

Index

output impedance
 (source follower), 249
voltage amplification
 (common source),
 247–248
voltage amplification
 (source follower),
 248–249
field strength
 electric, 10–11
 electromagnetic, 11
filament power demand, 263
flip-flop, 201–203
 asynchronous, 201
 J-K, 201, 202–203
 M-S, 202
 R-S, 201
 R-S-T, 203
 T, 203
flux
 lines of, 12
 magnetic, 12
flux density, 12, 177
Foot/Pound/Second System, 22
force, 18–19
forward current, 232
forward transconductance
 (common source), 246
Fourier series, 92–93
free-space wavelength, 268–269
frequency, 7–8, 151–154

frequency vs. wavelength, 268
functions, 100–107
 continuous, 104, 105
 cubic, 106
 discontinuous, 104, 105
 exponential, 107
 linear, 104
 logarithmic, 106–107
 nth-order, 106
 one-one, 100, 102
 one-to-one
  correspondence, 102
 onto, 102
 quadratic, 104–105
 quartic, 106
 trigonometric, 77, 107
fundamental units, 1–19

gain, 14–16
 current, 15
 power, 16
 voltage, 15
gauss per oersted, 14
general electron-tube circuit, 259
general mathematical
 symbols, 48–54
geometric interpolation, 301
geometric series, 89–90
Greek alphabet, 43–48
 lowercase, 45–47
 uppercase, 44–45
Greenwich meridian, 72

harmonic series, 90
Hay bridge, 288, 290
henry, 8
hertz, 7–8
hexadecimal numbers, 189
higher-order derivative, 117
highpass filters, 213–217
    constant-k, 213–214
    series m-derived, 214–216
    shunt m-derived, 216–217
hole, 4
hybrid parameters (common base), 243–244
hybrid parameters (common collector), 244–245
hybrid parameters (common emitter), 242–243
hysteresis loss, 184–185

imaginary numbers, 158
impedance, 9–10, 160–162
    absolute-value, 10
    complex, 9–10
indefinite integral, 124, 127–130
induced voltage, 179–180
inductance, 8
inductive reactance, 160
inductive susceptance, 163–164
inductor, losses in, 183–186

input capacitance, tube, 257
input power sensitivity, tube, 265
instantaneous amplitude, 155, 157
integrals, 121, 124–125, 127–133
integrator, 130–133
intersection, set, 98, 101
integration by parts, 127
interpolation
    arithmetic, 300–301
    geometric, 301
inverter (NOT gate), 196, 197

J-K flip-flop, 201, 202–203
j operator, 9, 158
joule, 4–5

kilogram, 2
Kirchhoff's law
    for DC amperage, 140, 141
    for DC voltage, 143, 144

latitude, 70–72
LC bridged T null network, 296–297
LC circuit, 206
linear acceleration, 18
linear function, 104
linearity, 125–126

lines of flux, 12
logarithmic function, 106–107
logarithmic series, 93–94
logarithms, 79, 82–84
    common, 82
    Napierian, 82
    natural, 82
log-log graph, 67
logic gates, 196–199
    AND, 196, 198
    inverter (NOT), 196, 197
    NAND, 197, 199
    NOR, 197, 199
    OR, 197, 198
    XOR, 197, 199
longitude, 70–72
loss, 14–16
    current, 15
    eddy-current, 184
    hysteresis, 184–185
    in matched lines, 278–279
    ohmic, 183
    power, 16
    SWR, 279–280
    voltage, 15
lowpass filters, 208–213
    constant-k, 209–210
    series m-derived, 210–211
    shunt m-derived, 211–213
luminous intensity, 3

Maclaurin series, 92
magnetic field intensity, 13
magnetic flux, 12
magnetic flux density, 12
magnetic pole strength, 13
magnetic unit conversions, 28, 34–35
magnetic units, 12–14
magnetizing force, 14
magnetomotive force, 13, 178–179
mass, 2
master, 202
master-slave (M-S) flip-flop, 202
mathematical notation, 43–63
Maxwell bridge, 289–291
measurement, 287–301
measurement error, 300
meter, 2
meter cubed, 16
meter per second, 17
meter per second per second, 18
meter per second squared, 18
meter squared, 16
meter squared per volt-second, 12
Meter/Kilogram/Second (MKS) System, 1–3
mho, 6, 163
miscellaneous unit conversions, 35, 36–40
miscellaneous units, 16–19

Index

mobility, 11
mole, 3
M-S flip-flop, 202

NAND gate, 197, 199
NAND operation, 195
Napierian logarithms, 82
natural logarithms, 82
negative peak amplitude, 156
newton, 18–19
noise, 227–230
noise figure, 230
NOR gate, 197, 199
NOR operation, 196
NOT gate, 196, 197
NOT operation, 194
null networks, 295–301
    LC bridged T, 296–297
    RC bridged T, 298
    RC parallel T, 299–300
numbering systems, 187–193

octal numbers, 189
oersted, 13, 14
ohm, 5, 8–9, 10, 144
ohmic loss, 183
ohm-meter, 5–6
Ohm's law
    for DC amperage, 138
    for DC resistance, 145
    for DC voltage, 142
one-one function, 100, 102

one-to-one correspondence, 102
onto function, 102
OR gate, 197, 198
OR operation, 195
ordinate, 67
output impedance (source follower), 249
output resistance (cathode follower), 256
Owen bridge, 291–292

partial sum, 87
peak amplitude, 156
peak-to-peak amplitude, 157
period, 7, 269–270
permeability, 14, 178
permittivity, 11
perveance
    diode, 252
    triode, 252, 253
phase, 151–154
phase angle, 157
    RC, 162
    RL, 161
plane angular measure, 17
plate amplification factor, 255–256
plate efficiency, 265
plate power dissipation, 265
polar axis, 71
polar coordinate plane, 68–70
pole strength, magnetic, 13

Index

positive peak amplitude, 156
positron, 4
potential difference, 5, 140
power, 7
    AC, 170–172
    apparent, 171–172
    DC, 146–147
    reactive, 171
    real, 170–171
power amplification and gain, tube, 263
power gain, 16
power loss, 16
power series, 90–91
precedence of operations, 63
prefix multipliers, 22, 23
progression, 86
proper subsets, 99
proton, 4
P:S turns ratio, 181

quadratic function, 104–105
quartic function, 106

radian, 17
radian per second, 18
radian per second per second, 18
radian per second squared, 18
radiation resistance, 281
range, function, 102–104

RC bridged T null network, 298
RC parallel T null network, 299–300
RC phase angle, 162
reactance, 8–9, 10, 160
    capacitive, 161
    inductive, 160
reactive energy, 173
reactive power, 171
real energy, 173–174
real power, 170–171
rectangular coordinates, 67
rectification efficiency, 233
reflection coefficient, 277–278
rel, 14
reluctance, 13–14, 175–177
reluctances in series, 176
reluctances in parallel, 176–177
required DC supply voltage, 258
required external cathode resistance, 261
required external plate resistance, 261
required external screen resistance, 261–262
resistance, 5, 6, 9, 10
    DC, 144–146
resistance parameters (common base), 239–240

Index

resistance parameters (common collector), 241–242
resistance parameters (common emitter), 240–241
resistivity, 5–6
resonant frequency, 205
resonant harmonic antenna, 284–285
resonant unterminated long wire, 285
RF transmission lines, 270–280
right ascension, 72
RL phase angle, 161
root-mean-square (rms) amplitude, 157
rounding, 61–62
R-S flip-flop, 201
R-S-T flip-flop, 203

Schering bridge, 293–294
scientific notation, 56–59
screen amplification factor, 256
screen current, 260
secant, 78
second, 2, 7
secondary binary operations, 195–196
second derivative, 117
semiconductors, 231–249
semilog graph, 67
sequence, 86
series, 86–88
    arithmetic, 89
    arithmetic-geometric, 91
    basic, 89–91
    conditionally convergent, 88
    convergent, 87
    divergent, 87
    exponential, 93–94
    Fourier, 92–93
    geometric, 89–90
    harmonic, 90
    logarithmic, 93–94
    Maclaurin, 92
    power, 90–91
    Taylor, 91–92
    trigonometric, 93
set, 97–101
    intersection, 98, 101
    union, 99, 101
SI unit conversions, 24–27
siemens, 6, 163
siemens per meter, 6
signal output power, 264
signal-plus-noise-to-noise ratio, 229
signal-to-noise ratio, 228–229
significant figures, 60–63
sine, 77
slave, 202
solid angular measure, 17

source-follower FET circuit, 248
S:P turns ratio, 181
spherical coordinates, 76–77
square meter, 16
Standard International System of Units, 1–3
standing-wave ratio, 275–280
    current, 276–277
    voltage, 276
static forward current transfer ratio, 235
static resistance, 232–233
steradian, 17
subscripts, 54–55
subsets, 99
superscripts, 55–56
susceptance
    capacitive, 164
    inductive, 163–164
susceptibility, electric, 11

T flip-flop, 203
tangent, 77
Taylor series, 91–92
temperature, 2–3
tesla, 12
tesla-meter per ampere, 14
theorems in algebra, 65–67
thermal noise power, 227–228
thermal noise voltage, 228
time, 2

transconductance, 255
transformer, 180–186
    current demand, 182–183
    efficiency, 181
    impedance transformation, 182
    losses in, 183–186
    P:S turns ratio, 181
    S:P turns ratio, 181
    voltage transformation, 181–182
transistor *see* bipolar transistor, field-effect transistor
transmission lines, RF, 270–280
transmission-line section
    1/4 wave, 207–208
    1/2 wave, 208
trigonometric/exponential formulas, 95
trigonometric functions, 77, 107
trigonometric identities, 79–82
trigonometric series, 93
trigonometry, 77–84
triode
    3/2 power for, 253
    perveance, 252, 253
truncation, 60–61
tubes, 251

union, set, 99, 101
unit circle, 83
unit electric charge, 4
units
    electrical, 4–12
    fundamental, 1–19
    magnetic, 12–14
    miscellaneous, 16–19
    Standard International (SI) System, 1–3
    Meter/Kilogram/Second (MKS) System, 1–3

valves, 251
vectors, 108–114
    cross product of, 112, 113
    dot product of, 112, 113
    in polar plane, 110–112
    in xy-plane, 108–109
    in xyz-space, 112–114
velocity
    angular, 18
    linear speed, 17
velocity factor, 271, 272
Venn diagrams, 100
vernal equinox, 72
volt, 140–141
volt rms, 168
volt per meter, 10–11
voltage
    AC, 168–170
    DC, 140–144
    induced, 179–180
voltage amplification (common source), 247–248
voltage amplification (source follower), 248–249
voltage amplification and gain, tube, 262
voltage gain, 15
voltage loss, 15
voltage standing-wave ratio, 276
volume, 16

watt, 7, 146
watt per square meter, 11
waveform derivatives, 121–124
waveform integrals, 130–133
wavelength
    electrical, 271–273
    free-space, 268–269
weber, 12
Wheatstone bridge, 294, 295
Wien bridge, 294, 296

XOR gate, 197, 199
XOR operation, 196
xy-plane, 67, 77
xyz-space, 73, 74

About the Editor in Chief

Stan Gibilisco has authored or coauthored dozens of nonfiction books about electronics and science. He first attracted attention with *Understanding Einstein's Theories of Relativity* (TAB Books, 1983). His *Encyclopedia of Electronics* (TAB Professional and Reference Books, 1985) and *Encyclopedia of Personal Computing* (McGraw-Hill, 1996) were annotated by the American Library Association as among the best reference volumes published in those years. Stan's work has gained reading audiences in the Far East, Europe, and South America. He maintains a Web site at

http://members.aol.com/stangib